Integrated Circuits and Systems

Series editor

Anantha P. Chandrakasan, Massachusetts Institute of Technology
Cambridge, MA, USA

More information about this series at http://www.springer.com/series/7236

Hideto Hidaka
Editor

Embedded Flash Memory for Embedded Systems: Technology, Design for Sub-systems, and Innovations

 Springer

Editor
Hideto Hidaka
Renesas Electronics Corporation
Tokyo
Japan

ISSN 1558-9412
Integrated Circuits and Systems
ISBN 978-3-319-85624-7   ISBN 978-3-319-55306-1   (eBook)
DOI 10.1007/978-3-319-55306-1

Printed on acid-free paper

This Springer imprint is published by Springer Nature
The registered company is Springer International Publishing AG
The registered company address is: Gewerbestrasse 11, 6330 Cham, Switzerland

Contents

1 Introduction.. 1
Hideto Hidaka

2 Applications and Technology Trend in Embedded
Flash Memory.. 7
Hideto Hidaka

3 Overview of Embedded Flash Memory Technology.............. 29
Takashi Kono, Tomoya Saito and Tadaaki Yamauchi

4 Floating-Gate 1Tr-NOR eFlash Memory..................... 75
Antonino Conte, Fabio Disegni, Francesco La Rosa
and Alfonso Maurelli

5 Split-Gate Floating Poly SuperFlash® Memory Technology,
Design, and Reliability 131
Nhan Do, Hieu Van Tran, Alex Kotov and Vipin Tiwari

6 SONOS 1Tr eFlash Memory 179
Hidenori Mitani and Ken Matsubara

7 SONOS Split-Gate eFlash Memory 209
Takashi Ito and Yasuhiko Taito

Index 245

Contributors

Antonino Conte Microcontrollers and Digital ICs Group, STMicroelectronics, Catania, CT, Italy

Fabio Disegni Automotive and Discrete Group, STMicroelectronics, Agrate Brianza, MB, Italy

Nhan Do Microchip Technology Inc, Chandler, USA

Hideto Hidaka Renesas Electronics Corporation, Tokyo, Japan

Takashi Ito Core Technology Business Division, Renesas Electronics, Kodaira-shi, Tokyo, Japan

Takashi Kono Core Technology Business Division, Renesas Electronics, Kodaira-shi, Tokyo, Japan

Alex Kotov Microchip Technology Inc, Chandler, USA

Ken Matsubara Core Technology Business Division, Renesas Electronics, Kodaira-shi, Tokyo, Japan

Alfonso Maurelli Automotive and Discrete Group, STMicroelectronics, Agrate Brianza, MB, Italy

Hidenori Mitani Core Technology Business Division, Renesas Electronics, Kodaira-shi, Tokyo, Japan

Francesco La Rosa Microcontrollers and Digital ICs Group, STMicroelectronics, Rousset Cedex, France

Tomoya Saito Core Technology Business Division, Renesas Electronics, Kodaira-shi, Tokyo, Japan

Yasuhiko Taito Core Technology Business Division, Renesas Electronics, Kodaira-shi, Tokyo, Japan

Vipin Tiwari Microchip Technology Inc, Chandler, USA

Hieu Van Tran Microchip Technology Inc, Chandler, USA

Tadaaki Yamauchi Renesas Electronics, Kodaira-shi, Tokyo, Japan

Chapter 1
Introduction

Hideto Hidaka

An embedded system is defined as "a computer system with a dedicated function within a larger mechanical or electrical system, often with real-time computing constraints, embedded as part of a complete device often including hardware and mechanical parts." It realizes far better cost/performance by specialized hardware and software with minimal amount of memory and CPU resources for real-time control, often provided by a one-chip MCU (Micro-Controller Unit) than general-purpose computing systems (Fig. 1.1). Today embedded systems are applied to almost all of the equipment with electrical controls in consumer products, power modules, motor-control module, medical equipment, automotive engine control units, etc. They will increasingly play important roles in future smart/mobile society, particularly in smart mobile phones, automobiles, smart factory, and IoT/IoE (Internet of Things/Everything) systems as their essential elements for enabling smarter and intelligent systems. Scales in numbers and pervasiveness will grow according to the demands for sophisticated controls of things in the everyday life of humans and factories enabled by pervasive sensors and computing power to feed data for advanced judgment and feedback systems.

Figure 1.2 describes an early prototype of a human body-attached thin sensory computing node with integration <1 inch-square for 200 MHz internal operations [1]. Equipped with 9 dimensional sensors for position/velocity/acceleration ($\times 3$ dimensions) and Bluetooth® low energy communication, a 200-MHz MCU with 2-MB fast embedded flash memory provides a very fast process as well as very low power for battery operations to provide all of the necessary data-processing power and latency for human body-attached sports applications, thus achieving an unprecedented power efficiency of 20-μA/MHz operation @200 MHz. As essential components in embedded systems, MCUs will continue to play important roles to define embedded-system properties such as performance, power, form factor, etc.

H. Hidaka (✉)
Renesas Electronics Corporation, Tokyo, Japan
e-mail: hideto.hidaka.pz@renesas.com

© Springer International Publishing AG 2018
H. Hidaka (ed.), *Embedded Flash Memory for Embedded Systems: Technology, Design for Sub-systems, and Innovations*, Integrated Circuits and Systems, DOI 10.1007/978-3-319-55306-1_1

1

Dedicated H/W and S/W optimized for single-purpose use for:

- Power/performance/cost optimized for specific uses
- Real-time computing tasks supported by small controllers
  (micro-controller units)
- Compact optimized integration of H/W and S/W for
  power efficiency and small form factors

Fig. 1.1 Core advantages of embedded systems

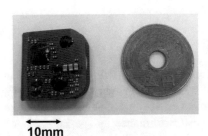

- 9D-sensors
- Bluetooth® Low Energy
- 200MHz 4mA@1.1V
  (20uA/MHz @200MHz operation)
- 2MB/64KB-ROM(for code/data),
  128KB-RAM with 2KB-Cache
- 40nm eFlash process

10mm

Fig. 1.2 An example of embedded system for an IoT sensory node [1]

Embedded systems with embedded components for small form factors contribute well to edge-located sensory nodes and terminals in the IoT age. Smart mobile phones not only integrate all of the necessary functions of everyday human life but also have come to provide edge devices for cloud systems to provide cyber-physical systems in all. IoT/IoE applications will develop to cover human motions/lives/health as well as motors and cars. Embedded physical-mobility applications in automotive and robot controls incorporate and integrate human- to automobile-level dynamic control together with information processing. Real-time sensing and feedback is the key enabling factor for advanced automotive control for ADAS (Advanced Driver Assistance System). This will open up new societal and welfare approaches. These most advanced controls begin to employ learning and reasoning for judgments with critical signal/data/information disseminated in society, necessitating embedded security.

Major factors of core technologies to support such a future smart/mobile society are security, safety, communication, signal sensing and processing, and low power to be embedded in real-time edge computing schemes under large-scale system designs. These are also essential elements in future embedded systems. Embedded-system designs possess a number of enabling elements including embedded memory evolution and revolution, low-power designs in technology, circuit and systems, and 2.5D/3D integrations such as chip stacks, etc. Among these factors, embedded memory is by far the most important basic technology to define system cost, performance, and power. Embedded memory, defined as "memory function with memory interface embedded inside a chip, not exposed to the chip interface," is now dominated by embedded SRAM and embedded flash memory in the market with possible introduction of emerging non-volatile memory such as magnetic-RAM and oxide-RAM in the future. They are mostly intended for work memory—such as RAM for data/code memory and ROM for code storage—often in combination with stand-alone data storage in RAM and ROM (SRAM, DRAM, NAND-flash, hard-disk drive, etc.) for memory hierarchy.

Since its inception in the early 1990s [2], embedded flash memory (eFlash) for MCU (micro-controller unit) products has realized revolutionary advancements by newly introduced programmable instruction code functions, which triggered the "eFlash innovation" age in the multi-billion dollar business of MCU segments in the semiconductor industry. Embedded flash memory has penetrated the market well to become an indispensable element in all MCU products for real-time embedded control applications. Starting at a 0.8-μm technology node in 1991 [2], eFlash technology has been developed in combination with standard CMOS logic technology for MCU to reach 28 nm of product level in 2015 [3], followed by the demonstration of eFlash cell operation by 16 nm-FinFET technology [4]. The eFlash innovation has its roots in cost reduction and value enhancement. By reducing the overall cost of hardware production, inventory, and embedded-systems development, field programmability also provides upgradable small embedded-systems features to support continuous market growth based on supply-chain innovations.

Advantageous properties of embedded flash memory come from programmability (for supply-chain innovation), non-volatility (for better power management and low power), and on-chip embeddedness (for compact optimized systems, etc.), which are particularly beneficial to embedded-systems designs (Fig. 1.3). These three factors have proved to be useful notably in automotive engine control and smart-card integration, which have actually driven the eFlash technology development and applied market. Programmable and updatable code storage for state-of-the-art engine control and secure data storage in smart-card applications are issues that eFlash has addressed most effectively and successfully in the history of the MCU industry. The most affordable small system on a chip, as realized by CPU and eFlash, is widely actualized into MCU products in which eFlash is an indispensable element in embedded-systems designs.

Embedded-system applications in automotive, smart IC cards, and low-power control have led the cutting-edge technology developments for eFlash so far and are now expanded by advanced embedded sub-systems such as ADAS, OTA (over-the-air S/W updates), embedded security platforms, and IoT/IoE requirements for smarter embedded applications. Future prospects of emerging non-volatile memory species for the new memory hierarchy, intended for lower-power and memory-bottleneck free embedded systems, are also investigated.

Embedded flash memory technology has its own unique history, quite apart from stand-alone flash memory counterparts for data-storage applications. Because of high-speed and low-power requirements in code-storage applications, split-gate memory cell structures are accepted only in embedded uses. Moreover, because of requirements in automotive applications that are quite different from stand-alone

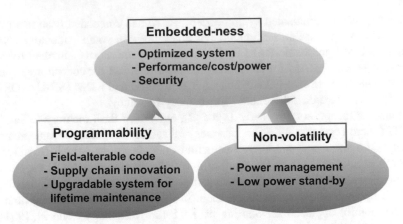

Fig. 1.3 Three types of advantages in embedded flash memory

products, it has partly exited from established floating-gate structures to employ breakthrough charge-trapping structures such as SONOS (poly-silicon–oxide–nitride–oxide–silicon) and nano-dot structures to provide higher reliability and lower-power eFlash development. Now eFlash technology development faces scalability challenges to be compatible with CMOS logic structures employing high-k metal gate and FinFET at the 28-nm technology node and beyond.

This book focuses on current mainstream eFlash technologies, in their origins, trends, and future prospects, from the viewpoints of embedded-systems designs. Representative cell structures are described each in 1Tr (transistor) and 1.5Tr (transistor) split-gate configurations for real applications and organized as shown in Fig. 1.4. It will not describe physical flash-memory device technologies in detail or in a comprehensive manner but will focus only on the mainstream eFlash technology in the market.

Chapter 2 (by Hideto Hidaka) introduces applications and requirements for embedded flash memory in the micro-controller unit (MCU) products and then describes how and why embedded flash memory has created supply-chain innovation (embedded flash innovation) as well as expanded the applied market by meeting the requirements supported by process, device, and circuit-technology evolutions. In the future projection, some of the requirements beyond the intrinsic limitations of eFlash and rationales for emerging non-volatile memories are discussed.

Chapter 3 (by Takashi Kono, Tomoya Saito, and Tadaaki Yamauchi) gives an overview of representative embedded flash-memory technologies to give an overall picture of the current mainstream embedded flash memories to conclude the validities of stacked floating-gate and SONOS charge-trapping structures as implemented in 1Tr and 1.5Tr split-gate cell configurations. This is followed by basic design concepts of embedded flash memories in memory-array architectures and operations as well as sub-system-level design concepts.

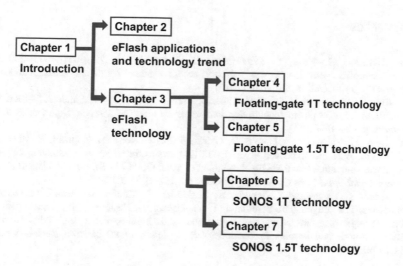

Fig. 1.4 The organization of this book

Each of those representative embedded flash-memory technologies is further discussed in Chaps. 4 through 7 with a detailed description of each mainstream embedded flash memory technology and design concept reviewed extensively to focus on embedded-specific flash-memory solutions.

Chapter 4 (by Antonino Conte, Fabio Disegni, Francesco La Rosa, and Alfonso Maurelli) is dedicated to floating-gate 1Tr cell technology and design, which is the most mature embedded flash-memory technology. Three illustrative descriptions of the concepts and implementations are described in security-intensive, power-sensitive, and automotive-grade designs.

Chapter 5 (by Nhan Do, Hieu Van Tran, Alex Kotov, and Vipin Tiwari) describes 1.5Tr split-gate cell technology and design based on a floating-gate structure. Original SuperFlash® memory technology, design, and reliability are discussed as is its development history and future projections.

Chapter 6 (by Hidenori Mitani and Ken Matsubara) is dedicated to SONOS 1Tr technology and design as well as the history of SONOS flash-memory development and future prospects. A low-cost, very low-power eFlash implementation is described and discussed for possible applications.

Chapter 7 (by Takashi Ito and Yasuhiko Taito) focuses on 1.5Tr cell technology and design. A high-performance eFlash technology for automotive-control applications is described in terms of design concepts and implementation for performance and reliability. Future prospects in scalability and technological challenges are also discussed.

References

1. M. Nakajima, I. Naka, F. Matsushima, T. Yamauchi, A 20 µA/MHz at 200 MHz microcontroller with low power memory access scheme for small sensing nodes. in *Proceedings of COOL Chips 2016* (2016)
2. C. Kuo, T. Toms, N.M. Weidner, H. Choe, D. Shum, K.-M. Chang, P. Smith, A 512 KB flash EEPROM for a 32 bit microcontroller. in *Symposium on VLSI Circuits, Digest of Technical Papers*, pp. 87–88 (1991)
3. Y. Taito, M. Nakano, H. Okimoto, D. Okada, T. Ito, T. Kono, K. Noguchi, H. Hidaka, T. Yamauchi, A 28 nm embedded SG-MONOS flash macro for automotive achieving 200 MHz read operation and 2.0 MB/s write throughput at Tj of 170 °C. in *Digest of Technical Papers, International Solid-State Circuits Conference*, pp. 132–133 (2015)
4. S. Tsuda, Y. Kawasima, K. Sonoda, A. Yoshitomi, T. Mihara, S. Narumi, M. Inoue, S. Nuranaka, T. Maruyama, T. Yamashita, Y. Yamaguchi, D. Hisamoto, First demonstration of FinFET split-gate MONOS for high-speed and highly-reliable embedded flash in 16/14 nm-node and beyond. in *Technical Digest—International Electron Devices Meeting*, 11.1 (2016)

Chapter 2
Applications and Technology Trend in Embedded Flash Memory

Hideto Hidaka

2.1 Embedded Flash Memory in MCU History

Compared with a stand-alone memory device, embedded memory is defined as a memory function with a memory interface embedded inside a chip and not exposed to the chip interface. In the present market, embedded SRAM is extensively utilized for working memory in MCU and SOC applications to comfortably match with the CMOS logic technology environment. In parallel, embedded flash-memory advantages listed in Fig. 2.1 provide critical solutions in embedded-systems designs. These advantageous properties of embedded flash memory come from programmability (for supply-chain innovation), non-volatility (for better power management and low power), and on-chip embeddedness (for compact optimized systems, etc.) and are particularly beneficial for embedded system designs. These three points will be discussed in Sect. 2.1–2.3.

A non-volatile memory in today's industry actually provides a memory to store information for >10 years while the power supply is turned off. Non-volatile memory has been preferred especially in remote local systems in eliminating the demands for power supply and battery back-up for system simplicity and maintainability. It also helps mitigate environmental concerns against the extensive use of batteries. Most of the non-volatile memories now in production are ROM and programmable ROM. A family of ROM technologies—including one-time programmable ROM (OTP ROM), electrically erasable and programmable ROM (EEPROM), and Flash Memory—utilizing charge (electron or hole) storage structure is classified in Fig. 2.2. The term "flash memory" implies fast erasure operations in blocks of memory cells to accelerate the erase operation in bulk. The distinction between programmable ROM and RAM lies in their performance. RAM

H. Hidaka (✉)
Renesas Electronics Corporation, Tokyo, Japan
e-mail: hideto.hidaka.pz@renesas.com

© Springer International Publishing AG 2018
H. Hidaka (ed.), *Embedded Flash Memory for Embedded Systems: Technology, Design for Sub-systems, and Innovations*, Integrated Circuits and Systems, DOI 10.1007/978-3-319-55306-1_2

- **Lower system cost**
- **Fast system development**
- **High performance code storage**
- **Reliability, security, safety**

- **CMOS-oriented integration**
- **Flexible selection of memory capacity**
- **Low power memory interface**

Fig. 2.1 Advantages of embedded flash memory in embedded systems

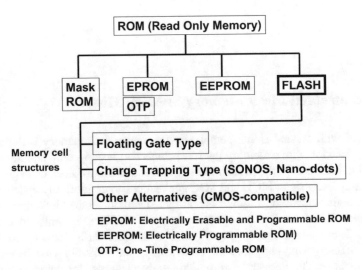

Fig. 2.2 Flash memory and related ROM devices

requires actually 10–20 years lifetime with rewrite capability with $>10^{16}$ cycles, whereas typically flash memories are limited to $<10^6$ cycles and much slower write operations (2–4 orders in time) because of the physical limitation in the SiO_x/SiN_y system.

In the history every ROM species listed in Fig. 2.3 has been embedded in the real products, initially for program updates after software development by OTP, and finally into the overall cost reduction by flash memory. Now flash memory has the second largest market size among embedded memory species, following only embedded SRAM. Presently the major application of embedded flash memory is for program-code storage in micro-controller (MCU) products, called "flash-MCU." Embedding flash memory on an MCU chip not only enhanced the values by embeddedness but innovated the supply chain to reduce the total cost of delivery.

MCU is an essential part in many of embedded systems to provide optimized compact system designs with real-time computing tasks by single-purpose embedded software implementation. Figure 2.3 describes the evolution of MCU products with regard to the use of embedded memory. After one-chip integration in 1970s, the embedded memory for program code storage in MCU evolved

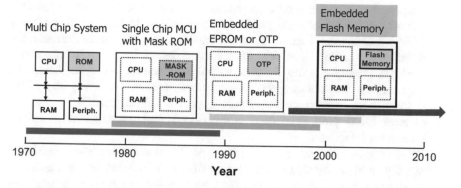

Fig. 2.3 Evolution of MCU by on-chip memory [1]

significantly every decade from mask ROM, OTP (one-time programmable ROM), and finally flash memory. In the 1990s, there were strong demands for applying the flash memory in MCU to save total cost and turn-around time of MCU system development. After the introduction of embedded flash memory in MCU, flash memory has improved its features such as optimized ROM features, single power-supply operation, and better reliability. The flash-MCU dominates the worldwide MCU market.

We have seen a great leap in the market penetration of embedded flash memory thanks to the overall cost reduction in the supply chain through design, production, and inventory control simply by programmability in flash-MCU. Here the overall cost advantage by the economy of scale exceeds the disadvantage of higher wafer-process cost for incorporating flash memory, which was a key factor for successful embedded flash-memory market development.

Integrating embedded flash memory in MCU products for programmable-code implementation has a large impact on the design and the supply chain for embedded systems as follows.

(1) Short turn-around time in system development: The embedded mask ROM scheme takes a longer time from the start of software development to the shipping of a system because program modifications and debugging require iterations of mask-ROM revisions to be shared by the system vendors and MCU vendors. In contrast, flash-MCU products are shipped with un-programmed (blank) states to customers during software development. On-chip flash memory is programmed after system assembly; hence, the developments of total system and program can proceed in parallel, thus resulting in a shortened turn-around time of system development. And the recovery time after debugging becomes shorter by recovering programs in the field. Therefore, the total cost and time of system development are reduced.

(2) Simplified and flexible production and inventory control: Variation of the final product sets is often offered by different firm-wares in the MCU market. If mask ROM MCU is used, each set must be manufactured with each different mask ROM data for MCU, which causes complicated inventory management. Flash MCU simplifies this situation by programming the MCU just before the set assembly, thus reducing the inventory cost. Likewise, at every step of the supply chain the inventory control with respect to different mask ROM data (for different MCU product models) is eliminated, thus causing a supply-chain innovation (Fig. 2.4).

(3) HW-SW separation and co-design: Flash-MCU has successfully separated HW and SW in the embedded system design where SW development is conducted independent of HW designs. In such a system, a further advantage emerges by HW-SW co-designs. For example, a lower-power system is realized by SW-programmable power management by programmable powering-up sequence compositions and use-case adaptable control of power-delivery schemes. The potential capability of software will advance the control scheme and create new values.

(4) Paradigm shift in software development by learning mechanism: In embedded-systems development by single-chip MCU, hardware control has been replaced by software control, and the product lineup is unified in hardware. Here a new problem has arisen in the turn-around time for software development becoming longer according to complex system requirements.

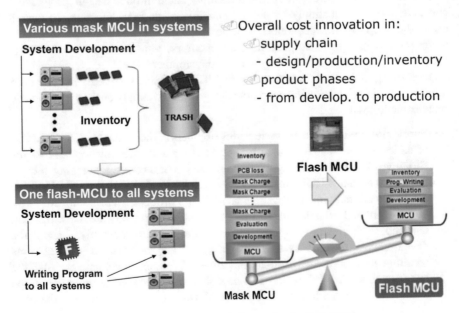

Fig. 2.4 Delivery-cost reduction in supply-chain innovation by flash-MCU

Flash MCU can alleviate some of this problem by introducing learning mechanisms for parameter updates in the system, thus providing new system values.

(5) Upgradable system for lifetime maintenance: A SW-upgradable system and lifetime maintenance scheme is enabled, thus providing another innovation in the supply chain. Especially the in-field SW update by over-the air (OTA) will contribute to actualizing such business models.

Embedded flash memory in MCU has achieved an innovation for increasing supply and demand based on the overall cost reduction in the supply chain and value creation for embedded-systems development. This is the basic rationale of embedded flash memory in MCU products for embedded systems. Technologically this hits the balance between application-specific optimization and minimum cost by programmability.

Together with the eFlash innovation in the basic cost structure previously described, finding new uses of eFlash—as well as developing eFlash technology and sub-system designs to form a standard and to meet new demands—has expanded the MCU market significantly. Figure 2.5 lists meaningful uses of eFlash in embedded systems. These are guiding concepts for eFlash applications in embedded-systems designs.

In addition to achieving values in the non-volatile storage of code and data, eFlash opened the door for design reform in wide levels of HW-SW co-designs to augment design efficiency in small embedded systems, security applications by using the hidden data on the chip, and upgradable VLSI functions. These are items to develop and fully utilize in future cyber-physical system (CPS) and IoT system designs. Finding new uses and applications of eFlash is still underway in every aspect of embedded-systems designs.

Fig. 2.5 Embedded flash-memory uses

- **Upgradable program codes**
  - **Flash innovation;**
    cost reduction in supply chain
  - **System boot, power on/off sequence etc.**
  - **OTA for maintenance &system update**
- **Data storage**
  - **Parameter settings   - data logging**
- **Selectable functions, re-configurable**
  - **FPGA configuration   - Function switches**
  - **Upgradable VLSI functionality**
- **Security   - Secure data/key storage**
- **Design flexibility in development**
  - **HW-SW partitioning and co-design**

2.2   Expanding Applications of Embedded Flash Memory

Unifying the technology and design into an established standard of flash-MCU configurations has contributed to widening the applications of eFlash. In addition to the program-code storage by embedded-systems program sizes (up to multiple megabytes), in many cases of embedded-systems designs small-capacity data storage (up to 32 kB) with multiple rewrite capability (100 K–1 M cycles) to replace the externally provided EEPROM is required. This one-chip code-data storage configuration of flash-MCU (Fig. 2.6) has been established to help increase the market volume effectively. This is supported by a reasonable combination of on-chip integration utilizing the same memory cell technology for both the code and data flash memory on the chip, which has been realized by technology development.

2.2.1   Standards for General Purpose MCU

In the typical flash-MCU architecture depicted in Fig. 2.7, the on-chip flash memory first merged code ROM with data ROM, thus providing a standard small-system configuration. In parallel, by developing applications for multiple-time programmability in real-time control and achieving an overall cost advantage, all the multi-billion dollar MCU markets now focus on flash-MCU solutions. Among them major segments of the market that have increased the market size significantly are automotive control and smart-card applications.

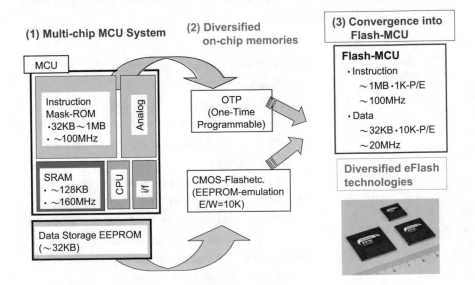

Fig. 2.6 Convergence into a standard flash-MCU configuration

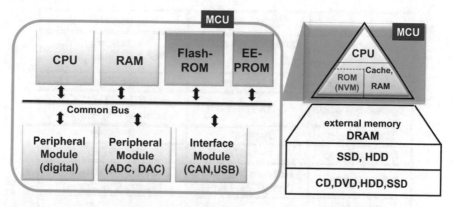

Fig. 2.7 Non-volatile memory in MCU architecture

Another important contributor to product convergence is the development of embedded-specific flash-memory technology attaining read performance competitive to mask-ROM configuration and high-temperature operation together with high reliability up to the automotive grade. The process and device technology for large-capacity stand-alone data memory such as NAND flash memory does not meet the requirements by embedded applications in read speed and reliability. These two distinct technology sets are shown in Fig. 2.8.

Accordingly, the technology map in Fig. 2.9 shows the discrepancy between stand-alone NAND flash memory and embedded flash memory in read and write performances due to requirements for fast data write in data-oriented stand-alone large-capacity memory and fast code read in embedded applications without frequent write.

The benefits of embedded flash memory come from its embedded nature, programmability, and non-volatility, as follows, compared with probable alternatives.

(1) Because in the embedded environment the off-chip memory access path and drivers are eliminated, high-speed and low-power operations are achieved, and data security is easy to implement with a less-accessible internal data bus. In addition, high-density physical packaging, enhanced reliability, reduced EMI, and lower system cost are provided by the embedded memory solutions.
It is to be noticed that 2.5D and 3D integration by stacked chips using TSV (Through Silicon Via) will provide alternatives approaching to embedded memory performance and power in large-capacity memory integrations.

(2) The flexibility of design in the embedded environment gives optimal design in memory capacity, memory interface configuration, functions, operating voltage, etc., for each application showing a tough challenge in optimizing and controlling the eFlash macro variations. Thus comes the eFlash compiler and technology/design platform approaches, now in wide usage in the industry.

An increasing number of function requirements in terms of safety, security, memory interface, etc., poses a new challenge in controlling the design and verification of the eFlash sub-system. High-level design and verification

	NAND Flash	Embedded Flash
Major Use	Data storage media	Code storage for real-time apps.
Memory Size	~16GByte	~16MByte
Operating Temp.	up to Tj=85°C	Auto: Tj=150 ~ 170°C Industry: Tj=125°C
Data Access	Sequential	Random
Random Read	~1MHz	≥ 100MHz
Data reliability	Need strong ECC (Long Delay time)	SEC/DED at most (Speed Constraint)
Compatibility w/ High Speed Logic CMOS	Not required	Mandatory

SEC : Single Error Correction
DED : Double Error Detection

Fig. 2.8 Embedded flash versus stand-alone NAND flash specifications [2]

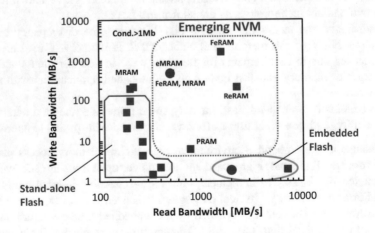

Fig. 2.9 Comparison of stand-alone and embedded flashes [3]

methodologies, such as HW-SW co-designs, will be essential in productivity management.

(3) The non-volatility attribute gives an opportunity for low-power design such as eliminating the battery-backup SRAM and achieving zero stand-by power by intermittent system operations. Attentions are focused on zero stand-by power memory because today's scaled embedded SRAM often sees a large stand-by current due to the scaled MOS FET in the memory cell.

Frequent backup of memory data to non-volatile storage necessitates great endurance as well as fast- and low-power rewrite operations at the flash memory. Emerging non-volatile memory properties are favored in intermittent operations.
(4) Field programmability will come into a new age of wireless-communicated program updates called OTA. Data-security functions and remedies for EMI reduction [2] will play important roles in the OTA era.

In contrast, the intrinsic disadvantages of embedded flash memory in terms of re-write power, large mask counts in production process, compatibility with CMOS logic process, and scalability are looming in some applications as is described in Sect. 2.3.

As listed in Fig. 2.10, embedded flash memory provides functions of code and data storage, backup storage, system boot, and trimming-information storage for memory and analog parts of the chip, etc. By applications we find code and data storage dominant in micro-controller unit (MCU), updatable coefficient parameter storage in DSP (digital-signal processor), data storage in smart IC cards and RF-D tags, and configuration storage in field-programmable gate array (FPGA) and re-configurable logic, etc. By uses in the stages from proto-typing to volume production, embedded flash memory acts as easy-to-change ROM storage for easily verifying system concepts, fixing program bugs, and supporting program updates as well as production and inventory control by unified product lineup.

Expanding applied products benefiting from previously mentioned standardization and convergence of eFlash configurations and benefits include the following.

- Micro-controller unit (MCU) for code and data storage
- Smart IC cards and RFID for secure data and key storage

● By functions
 - Code storage; system boot, user program, firmware, look-up table
 - Data storage; EEPROM emulation, shadow storage,
    frequently updated parameters and coefficients,
    state before power down
 - Add-on parameter storage for trimming information etc.

● By applications
 - MCU (micro controller unit); for code and data storage
 - DSP (digital signal processor); for coefficient storage
 - Smart-IC cards ;EEPROM data storage
 - RF-ID ; data storage
 - Reconfiguration register; FPGA(Field Programmable Gate Array) etc.

● By uses
 - proto-typing ; verify system concepts
 - system development ; program debug and updates
 - early productions ; program updates
 - volume productions ; production and inventory control
 - lifetime control of VLSI

Fig. 2.10 Embedded flash-memory applications

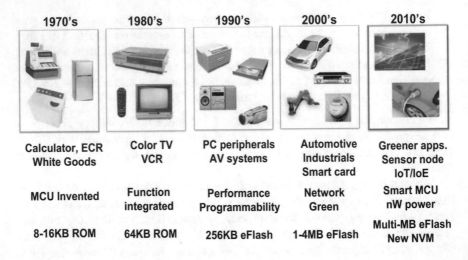

| 1970's | 1980's | 1990's | 2000's | 2010's |

Fig. 2.11 MCU market expanded by new applications

- Analog, power, FPGA, and SoC for add-on uses for power management, analog/RF circuit tuning, memory-redundancy program, re-configurations, etc. [4, 5].

In these established fields of applied products, demands for eFlash are increasing based on the growing number of software-upgradable engine control systems for minute controls of combustion as required by CO_x regulations, over-the-air (OTA) program upgrades, calibrations of analog circuitry, power management with programmable power-up sequences, security functions such as certifications, and learning functions for artificial intelligence processing, etc (Fig. 2.11).

2.2.2 Automotive Applications

In Sect. 2.2.2 and 2.2.3, representative flash-MCU applications in automotives and smart cards, which have significantly expanded the flash-MCU market since the 2000s, are described to illustrate how applications have driven the development of embedded flash-memory technology.

Automotive applications of flash-MCU illustrated in Fig. 2.12 indicate that today's electrically equipped car extensively uses MCUs for local real-time electrical control functions for engine control, body/chassis control, and peripheral functions, most of which employ embedded flash memory for locally storing the control programs, control parameters, and measured data. In modern automotive engine control, several sensors for crank angle, air-flow, and knocking phenomenon are connected to flash-MCU through application-specific ICs for pre-processing the sensed data. According to measured data from these sensors, the MCU gives feedback to control fuel injection, timing of the ignition plug and throttle motor,

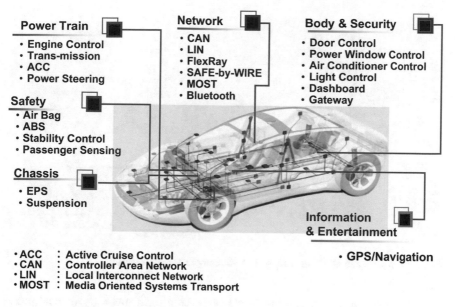

Power Train
- Engine Control
- Trans-mission
- ACC
- Power Steering

Safety
- Air Bag
- ABS
- Stability Control
- Passenger Sensing

Chassis
- EPS
- Suspension

Network
- CAN
- LIN
- FlexRay
- SAFE-by-WIRE
- MOST
- Bluetooth

Body & Security
- Door Control
- Power Window Control
- Air Conditioner Control
- Light Control
- Dashboard
- Gateway

Information & Entertainment
- GPS/Navigation

- ACC : Active Cruise Control
- CAN : Controller Area Network
- LIN : Local Interconnect Network
- MOST : Media Oriented Systems Transport

Fig. 2.12 Flash-MCU applications in a car

etc., to maintain engine operation under the most suitable conditions. Distributed real-time control like this in many embedded systems dominates MCU applications.

The trend in today's automotive control is strongly influenced by enhancing the combustion efficiency for environmental regulations, safety and security requirements, and getting more connective to outer information. These demands by new functions in a car add to uses of flash-MCU resulting in ≥ 200 units of flash-MCUs being used in a single car in some cases. By introducing electrical control in every part of the automotive control, even in combustion engine cars, and by introducing the IT region, the number of semiconductor parts used in a car amounts to 1–10 K. The electric car will use vast variety of flash-MCUs more extensively for motor control and associated power electronics.

One significant trend in modern cars is a merger of a conventional physical control regime for the engine and chassis and a newly introduced IT regime to connect with the outer cyber world to provide a higher level control and/or additional values in cars. Thus, "control meets IT" in a car (Fig. 2.13) where we are beginning to find a new space of cyber-physical systems to realize cognitive perception and automotive control [6]. Together with these functional advancements, once non-functional requirements such as safety and security have become essential elements in these embedded system designs as the system becomes more integrated and more connected. The Automotive Safety Integrity Level (ASIL) by ISO-26262 defines the safety and security requirements in automotive systems. ASIL-compliant MCU and SOC with secure IPs are becoming indispensable parts of automotive systems.

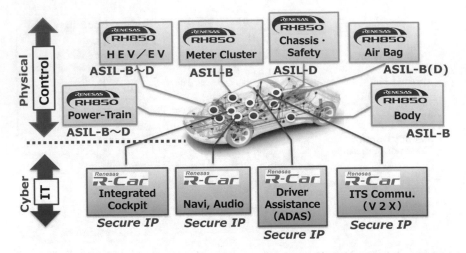

Fig. 2.13 MCUs and SOCs in a car with ASIL compliance and security

The performance and capacity trend of on-chip flash memory for automotive MCUs in Fig. 2.14 indicates ever-expanding technology requirements [2]. The overall MCU performance requirement has grown approximately by 20-fold in 10 years, 35% per year, which is supported by architecture evolution such as cache-memory usage and multi-core CPU implementation, eFlash speed enhancement, device scaling, design for reliability, etc. In parallel, the on-chip ROM capacity has grown by 23% per year to support the growth of on-chip code storage for

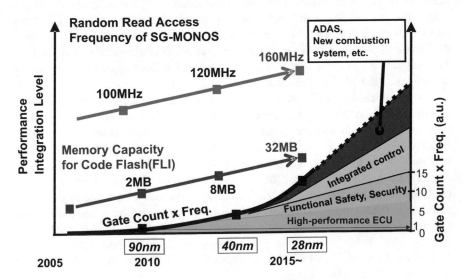

Fig. 2.14 Performance and ROM capacity of on-chip eFlash for automotive applications [2]

Fig. 2.15 Requirements for embedded flash memory in automotive uses

● **Memory band-width adapts to CPU speed**

● **Low power & low voltage operations**

● **Automotive-grade temperature operation and reliability**

    -40 to 150 C with very low failure rate

● **Offer overall supply chain cost lower than mask-ROM**

● **Safety and security functions**

● **CMOS logic embeddable technology**

program statements in automotive applications. In reality, MCU for automotive applications drives the development of scaled CMOS process integrated with embedded flash memory to achieve high-density memory as well as high-performance CMOS logic.

Examples of embedded flash-memory requirements for automotive applications in Fig. 2.15 reveal the actual challenges facing embedded flash-memory technology. From performance to match CPU to reliability under a very wide temperature range and low cost, embedded flash memory for MCU becomes the most challenging one in many aspects of today's semiconductor memory. In addition, safety and data security functions are becoming prevailing factors in the general requirements for MCU.

2.2.3 Smart-Card Applications

Smart cards present another application area where the capability of Flash-MCU has greatly expanded. Figure 2.16 shows an example of smart-card application in charging and paying for public transportation based on contactless card storage.

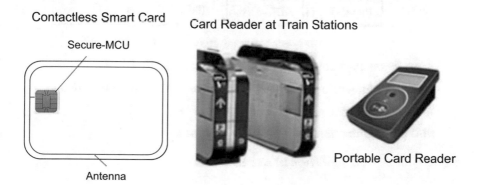

Fig. 2.16 Smart-card and applied systems [7]

Used in many everyday chip-card applications for money transactions, IDs, and data/configuration transport and storage in mobile phones, specially designed one-chip MCU called "secure-MCU" incorporates one-chip MCU with embedded EEPROM storage enclosed in logical/physical security protections. It is based on a security-intensive EEPROM with CPU or accelerator to process encryption/decryption in cryptography standards such as AES and RSA.

Embeddedness provides an advantage in data security because the memory interface is embedded on the chip, and tamper-resistant measures are easily implemented over the chip including the memory. The highest requirement for program/erase cycle is approximately 500 K-cycles, much higher than normal flash memory storage applications, because in many cases at every transaction the stored data as well as stored security key data are updated for higher security level. Thus, encrypted data and associated security keys are stored in a non-volatile memory array enclosed by security measures both logically and physically.

In the actual smart-card implementation (Fig. 2.17), near-field communication (NFC) is often adopted to provide the communication as well as power supply through a short-range wireless connection to the outer environment. Because the power-supply capacity by NFC is very limited, the current consumption by secure-MCU is required at <2 mA (max). Thus, peak power consumption is limited particularly in program/erase operations of the embedded-flash ROM.

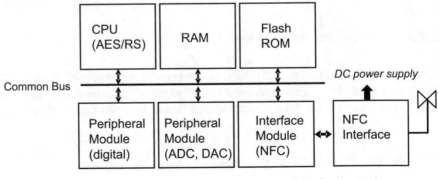

-Security encryption/decryption functions: AES/RS

-EEPROMwithverylow-powerprogram/erase;500Kcycles,1mA@peak

-Tamper resistance measures in hardware

-NFC communication/power-supply interface

Fig. 2.17 Secure-MCU organization for contactless smart card

2.2.4 Summary of Product Requirements

Figure 2.18 summarizes the maximum requirements for embedded flash memory in various MCU applications. From high performance under high temperature, as required in stringent automotive and industry conditions, to macro-area efficiency for small memory capacity for cost-sensitive applications in PC/OA and for the consumer, the requirements vary significantly. Particularly varied are the requirements for the memory capacity at 16 KB–8 MB, program/erase endurance at 1K–500 K cycles, and read performance from 10 MHz to 200 MHz inevitably force a number of optimal technology/circuit/sub-system designs to best fit the varied market. Even only in one technology we must offer a wide variety of ROM/RAM capacity, which is a basic parameter for MCU product line-ups to be efficiently deployed in a platform manner.

In Sect. 2.1 and 2.2, we have described how the favorable properties of embedded flash memory—such as embeddedness, programmability, and non-volatility—have contributed to innovative embedded-systems solutions. From prototyping uses to overall cost reduction for code storage in MCU, real-time control parameter updates, and security data memory, the Flash-MCU market has been expanding steadily. It also has found new market drivers in automobiles and smart cards. All together, flash-MCU has won one of the most successful businesses in embedded memory applications, second only to embedded SRAM in terms of market size. In the next Sect. 2.3, the future prospects of eFlash and new demands to be met by emerging non-volatile memory in the future will be overviewed.

		Automobile		Industry	PC/OA	Consumer	Secure-MCU
		Power Train	Body				
MCU	Performance (max. freq.)	~300MHz	150~200MHz	~300MHz	25~50MHz	20~100MHz	10~50MHz
	Power	0.5mA /MHz	0.5mA /MHz	1mA /MHz	0.5mA /MHz	0.25mA /MHz	0.1mA /MHz
	Temp.(Ta)	- 40 ~ 125 C		max 85 C	- 20 ~ 85 C		
FLASH	Density (max.)	8MB	2MB	1MB	2MB	1MB	1MB
	P/E cycle	Program Area : 1K / Data Area : 100K					100K-500K (EEPROM)
	Small Cell	✔					
	Small Macro				✔	✔	✔
	Fast Access	✔		✔			

Fig. 2.18 Requirements for flash-MCU technology for different applications with important factors in each application marked

2.3 Challenges and Prospects of Embedded Flash Memory in Embedded Systems

We have observed the major trend in embedded flash-memory technology for MCU applications as driven by current application requirements:

(1) Convergence into standard flash-MCU and add-on solutions.

  – code/data combination in design platform
  – add-on eFlash solutions.

(2) Multiple eFlash technologies are converging into selected species for device scalability and for low-cost eFlash macro, seeking for:.

  – basic memory cell scalability
  – integration compatible with advanced CMOS such as high-k metal gate, FD-SOI, and FinFET.

(3) Embedded flash memory design incorporates multiple functions:

  – low-power compact eFlash macro
  – byte-wide access, data security and safety functions, EMI reduction, etc.

Although the MCU market has converged into standard designs in flash-MCU solutions, other market segments—such as SoC, analog, and power products—are beginning to incorporate simple and small-capacity add-on flash memory for chip-ID and security functions etc. [4, 5]. Further convergence into scalable technologies and new standards for on-chip IP designs are now under investigations.

In contrast, some recent requirements by application trends exceed the intrinsic limitations of flash-memory properties.

(1) Slow and power-consuming in program/erase operation with limited numbers of endurance, flash memory is not suitable for applications with frequent data re-write such as driving recorder applications etc.
(2) Embedded flash-memory process costs relatively high in CMOS integrations and is often not compatible with underlying CMOS technology.

Hereafter in Sect. 2.3 some of future requirements to eFlash are discussed—specifically regarding functionality limitations, scalability, power consumption, and future system requirements—in Cyber-Physical System (CPS) designs.

Flash-memory technologies for embedded uses have come to deviate from stand-alone NAND flash-memory technology significantly (Figs. 2.8 and 2.9) because of the embedded-specific requirements such as host-logic CMOS compatibility, performance, lower cost with smaller capacity of on-chip memory, and reliability quite different from those for stand-alone data flash memory. For instance, multi-level storage, now commonly used in NAND flash memory, does not offer reasonable embedded solutions because the data reliability is degraded, and timing/area-consuming strong ECC cannot be accommodated in embedded uses for low-cost, small-capacity memory solutions. Diversified flash memory technology may require universal technology solutions arising from emerging non-volatile technologies as shown in Fig. 2.19. These candidates include ferro-electric polarization (FeRAM), magnetization orientation (MRAM), amorphous/crystalline phase change (PCRAM), filament formation by atomic motion (ReRAM), carbon-nanotube switch (NRAM), etc. In terms of functionality, the device properties of emerging non-volatile memories fill the gap between current ROM and RAM as well as that between stand-alone and embedded flash memories, thus not only partly unifying the technology but also ushering in new memory applications.

The scalability issue in eFlash is mostly attributed to the compatibility with the underlying advanced CMOS structure. New structures—such as high-k metal gate, FD-SOI, and FinFET—will have an impact on eFlash structures and CMOS compatibility. Thin-film storage structures—such as SONOS, nano-dot, and thinned floating gate—seem advantageous in technology nodes beyond 28 nm. In contrast, back-end integration in emerging memory, such as MRAM and resistive-oxide-RAM, will make the structure almost free from the underlying CMOS device structures and will find advantages if they realize the resistance against the thermal treatment by metal/inter-dielectric layer formations.

Significantly some of the emerging memories offer much improved performance over conventional flash memory in re-write operations as shown in Fig. 2.20. Achieving faster speed by 2–4 orders under lower voltage in re-write operations, these provide 2–3 orders lower energy than conventional flash memory. In addition,

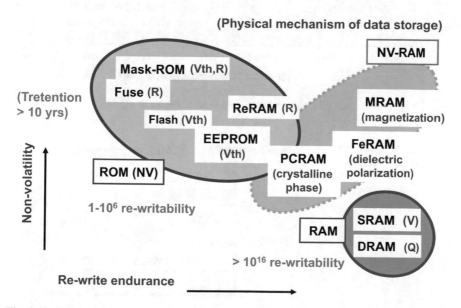

Fig. 2.19 ROM and RAM mapped with emerging non-volatile memory

they may provide lower cost because of simpler device structures. A key question remains as to whether these new physical memories will come into reality for embedded uses led by major innovation items like flash-MCU innovation in the IoT/IoE and AI era.

Memory hierarchy was originally intended for memory system cost and performance optimization. Recently it is seeking for lower-power systems, which has attracted great attention in computing system designs. In lower-tier memory hierarchy with almost-always stand-by state, non-volatile memory contributes to low stand-by power, and it can be also used in higher-tier memory hierarchies with intermittent power control for power-saving as shown in Fig. 2.21. Power-constrained designs in embedded systems for smaller-form factors and longer battery life in IoT-sensing nodes are expected to greatly benefit from this

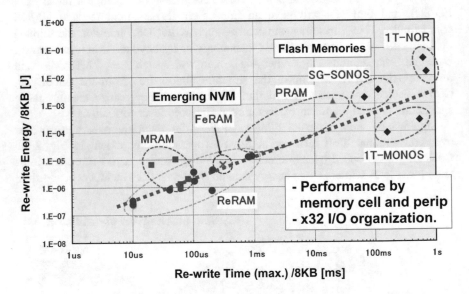

Fig. 2.20 Re-write energy in flash and other non-volatile memories [1]

Fig. 2.21 Power-saving by intermittent operations

dynamic power scaling by intermittent power control. In terms of cost, CMOS-compatibility, and re-write performance (speed, power, and re-write endurance), emerging non-volatile memories provide attractive solutions in future highly-integrated systems.

In modern moving apparatus—such as automobiles, drones, etc.—we often see requirements for merged control and information systems, i.e., cyber-physical systems (CPS). CPS designs have several essential points in common to many fields of applications such as automotive, industry, and IoT/IoE. Emerging mobility requirements in ADAS include safety, security, connectivity, low power with advanced sensing and feedback capability, which have impacts on the requirements for eFlash [6]. Together with conventional sensor-to-reaction feedback, we see a vast range of feedback control in multiple-path feedback loops as depicted in Fig. 2.22. From short-term fast memory to long-term, low-power memory adapting to learning mechanisms, embedded non-volatile memory will contribute to many aspects of CPS designs. Non-volatile memory will play important roles in every aspect of CPS designs in locally and asynchronously powered distributed systems including quite responsive feedback to long-term, learning-based intelligence.

Figure 2.23 summarizes a history of memory-centric VLSI evolution realized by on-chip non-volatile memory, which covers programmable product functionality and VLSI design methodology to market creation and supply-chain innovation. Future development of applications by static/dynamic power scaling, advanced learning, functional safety, security, OTA, and lifetime management of VLSI systems will certainly provide the source of next innovation in the IoT/IoE era.

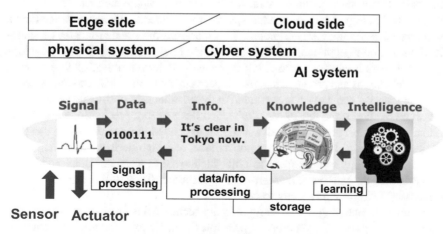

Fig. 2.22 Multiple-path feedback loops in modern control systems

	Innovation	Main Enabler	Product	Effect
1	Memory-based Logic	ROM Program Register-based computing	MPU, MCU	Program-driven Logic
				general-purpose logic
2	Alterable Program	SRAM/Flash	Flash-MCU FPGA	Re-configurable Supply chain innovation
				HW-SW co-design
3	Quasi-NVRAM Universal Mem.	NVM w/i energy-efficient re-write: (FeRAM, MRAM, ReRAM, NRAM...)	Unified-Memory/ MCU, SOC	Intermittent op. Re-usable logic Life-time management
				Learning, LP, FUSA/Security, OTA

Fig. 2.23 On-chip non-volatile memory triggers VLSI innovation [1]

2.4 Conclusion

In this chapter, we reviewed the history of embedded flash memory in MCU products in terms of supply-chain innovation, technology development, standardized products, expanding markets, and prospects for future technology and applications.

Embedded flash-memory technology originally inherited much from stand-alone flash-memory devices such as NOR flash memory. However, it has deviated from the main-stream stand-alone data flash-memory technology such as NAND flash memory technology due to embedded-specific requirements, and has advanced in its own path of technology development. Together with the standardization of specifications and configurations, dedicated technology has driven application development and market creation by making embedded flash memory adaptable to very broad applications and products in terms of performance, reliability, and overall cost. The rapid market penetration is attributed to the advantage in cost/value achieved by eFlash innovation since the 1990s. Embedded flash-memory technology and MCU market segment creations have advanced and expanded in interactive ways to proliferate by virtue of lower power, compact size, and high-performance integration with high reliability to contribute to embedded-systems advancements, notably in newly created automotive and smart-card applications. They will continue to serve humankind well for many years to come.

In contrast, emerging non-volatile memory technologies have been explored by introducing physical memory principles different from conventional silicon/silicon-oxide systems—such as ferroelectricity, magnetism, phase-change materials, and atomic motions—some of which have been productized in limited volumes. Although they are expected to partly solve flash-specific problems in terms of cost scalability, power consumption, and technological convergence for coming IoT/IoE and advanced mobility age, they have yet to prove real scalability compared with embedded flash memory, the ability to replace embedded flash memory and embedded SRAM in functions and performance, and the creation of new use cases and applications before coming into a "really emerged" species of semiconductor memories. Technologically they still suffer from insufficient properties under higher temperature in data retention and re-write endurance, thus limiting their scalability to a great degree. Basic research and development works are underway for solving these problems toward the next convergence in embedded non-volatile memory technology, and hopefully achieving the next innovation. This is a big challenge for humankind.

References

1. H. Hidaka, "Evolution of embedded flash memory technology for MCU," invited paper in Proceedings of the IEEE international conference on IC design and technology (2011)
2. Y. Taito, T. Kono, M. Nakano, T. Saito, T. Ito, K. Noguchi, H. Hidaka, T. Yamauchi, A 28 nm embedded split-gate MONOS (SG-MONOS) flash macro for automotive achieving 6.4 GB/s read throughput by 200 MHz no-wait read operation and 2.0 MB/s write throughput at T_j of 170 °C. IEEE J. Solid-State Circ **51**(1), 213–221 (2016)
3. Press-kit, *International Solid-State Circuits Conference*, p. 111 (2015), http://isscc.org/doc/2016/ISSCC2016_PressKit.pdf. 5 Nov 2015
4. C.C.-H. Hsu, Y.-T. Lin, E. C.-S. Yang, R. S.-J Shen, Logic non-volatile memory (World Scientific, Singapore, 2014)
5. H. Mitani, K. Matsubara, H. Yoshida, T. Hashimoto, H. Yamakoshi, S. Abe, T. Kono, Y. Taito, T. Ito, T. Kurafuji, K. Noguchi, H. Hidaka, and T. Yamauchi, A 90 nm embedded 1T-MONOS flash macro for automotive applications with 0.07 mJ/8kB rewrite energy and endurance over 100 M cycles under T_j of 175 °C in Technical digest international solid-state circuits conference, pp. 140–142 (2016)
6. H. Hidaka, How future mobility meets IT: Cyber-physical system designs re-visit semiconductor technology, Plenary Talk 3, Proceedings of the Asian solid-state circuits conference (2015)
7. http://www.smartcardalliance.org/smart-cards-intro-primer/. 30 Aug 2016

Chapter 3
Overview of Embedded Flash Memory Technology

Takashi Kono, Tomoya Saito and Tadaaki Yamauchi

3.1 Basic Components of an Embedded Flash System

In general, embedded non-volatile memory (eNVM) sub-systems in actual products are composed in a hierarchical manner. Figure 3.1 describes a typical embedded flash (eFlash) system as an example of eNVM systems. Here, the design of an eFlash system is categorized into three levels. Level 1 is related to memory cells. A suitable memory cell should be carefully chosen because it dominantly defines the total electrical characteristics and data reliability. Level 2 focuses on the peripheral circuit design of eFlash hard macros, which include such critical circuits as sense amplifiers, high-voltage generator, and so on. Electrical characteristics and data reliability of an eFlash system can be strongly dependent on the quality of these critical circuits. Level 3 includes the design techniques on an eFlash system level. Each functional block in the system shares proper roles to achieve system target specifications. In the following sections, the details of eFlash cell and circuit technologies in each level are overviewed.

3.2 Embedded Flash-Memory Cell Technologies

Figure 3.2 shows the category map of embedded programmable ROMs as an introduction of detailed explanation about "level-1." Here, memory capacity and the program/erase cycle (endurance) are used to categorize them. In addition, there are

T. Kono (✉) · T. Saito
Core Technology Business Division, Renesas Electronics, 5-20-1, Josuihon-cho, Kodaira-shi, Tokyo 187-8588, Japan
e-mail: takashi.kono.fv@renesas.com

T. Yamauchi
Renesas Electronics, 5-20-1, Josuihon-cho, Kodaira-shi, Tokyo 187-8588, Japan

© Springer International Publishing AG 2018
H. Hidaka (ed.), *Embedded Flash Memory for Embedded Systems: Technology, Design for Sub-systems, and Innovations*, Integrated Circuits and Systems, DOI 10.1007/978-3-319-55306-1_3

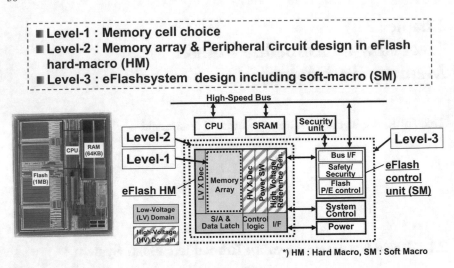

■ **Level-1 : Memory cell choice**
■ **Level-2 : Memory array & Peripheral circuit design in eFlash hard-macro (HM)**
■ **Level-3 : eFlashsystem design including soft-macro (SM)**

*) HM : Hard Macro, SM : Soft Macro

Fig. 3.1 A hierarchical approach in eFlash design

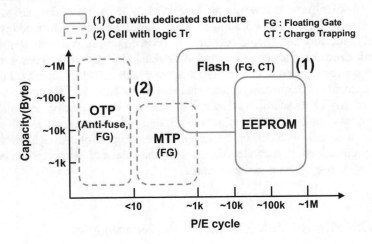

Fig. 3.2 Category of embedded programmable ROM

two groups according to the transistor structure used in a memory cell. The first group includes the memory cells with dedicated structure different from that of logic transistors provided in the base logic CMOS process. By introducing the "special" storage layer or structure in a cell, the memory cells in the first group (1) can provide attractive features in terms of cell size, performance, and reliability. Especially, a flash-memory cell is suitable for higher density by properly sharing diffusion node among cells. In contrast, they need additional masks to form memory cells, thus resulting in higher process cost. The memory cells in the second group (2) consist of logic transistors equipped in the base logic CMOS process. Typical examples are one-time programmable (OTP) memory cells and multiple-time programmable (MTP) memory cells. They can be integrated into base logic CMOS process at lower process cost thanks to no mask adder. On the contrary, they tend to be inferior to those in the first group in terms of cell size and reliability.

In this section, typical eFlash-memory cells are overviewed in terms of their cell structures, operation principle, and features in terms of characteristics and reliability, which is then followed by an explanation of other memory cells used in EEPROM, OTP/MTP, and CMOS logic flash.

3.2.1 History and Category of Embedded Flash-Memory Cells

Figure 3.3 shows an evolution tree of eFlash-memory cells for stand-alone and embedded uses. For stand-alone uses, such as NAND, smaller cell size is important. In contrast, reliability and performance take first priority for embedded uses. As a result, the memory cells for embedded uses have evolved on their own and have become diversified. In this tree, memory cells are categorized into two types from the viewpoint of storage layer structure: floating-gate structures (FG) and charge-trapping structures (CT). Each structure has been widely used in products and investigated intensively for many years.

Aside from the storage layer structure, the number of transistors in a memory cell (N_{tr}) is also an important factor that characterizes the memory cell. In most cases, N_{tr} is 1, 1.5 (generally called "split-gate"), and 2.

One-transistor (Tr) cell structure ($N_{tr} = 1$) has advantages in terms of cell size and density. Therefore, stand-alone flash memories with large capacity, such as NAND, generally adopt the 1Tr cell structure. However, the 1Tr cell structure is not

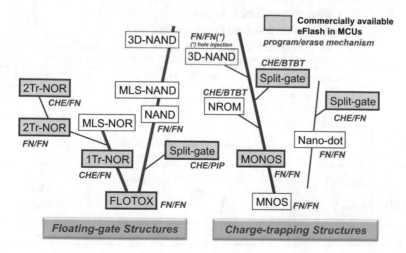

Fig. 3.3 Evolution and variation of eFlash-memory cells

always the best solution in embedded uses due to the following reasons. First, the pursuit of minimizing cell size in 1Tr cell structure often requires additional FEOL/BEOL processes [1]. Considering the fact that the area ratio of eFlash-memory array in a chip is not very high (e.g., approximately 10%), this cost adder for smaller cell size is not accepted in many cases. Second, the motivation to embed flash-memory array with logic circuits in place of externally connected flash memories is done to achieve higher performance, higher data reliability, and/or lower power consumption. Memory-cell structures other than the 1Tr memory cell have a chance to be adopted for embedded uses if they can meet such requirements.

A 1.5Tr or split-gate cell structure ($N_{tr} = 1.5$) has advantages compared with 1Tr cell structure in terms of performance. The first advantage is highly efficient programming. With the control gate (CG) biased to the voltage around the V_{th}, channel hot electrons can be efficiently generated at the region between the CG and the memory gate (MG) and injected into the charge storage layer under MG by a strong vertical electric field. This is called "source-side injection." Owing to its high efficiency, fast programming with very small cell current is achievable. The second advantage is that the CG of unselected cells (generally biased at 0 V) cut off the current leakage even when the V_{th} of erased cells is <0 V. This means that highly boosted CG level is not needed to secure sufficient cell current in read operation. In addition, no high voltages are applied to CGs and bit lines (BLs) even during program or erase operations of 1.5Tr cells. As a result, a fast read path can be built using only fast low-voltage logic transistors. A 2Tr cell structure ($N_{tr} = 2$) also has the latter advantage, but it lacks the source-side injection capability (Fig. 3.4).

From the viewpoint of read disturb immunity, it is ideal that memory gates (MGs) are always fixed at 0 V during read operations (assuming that source and

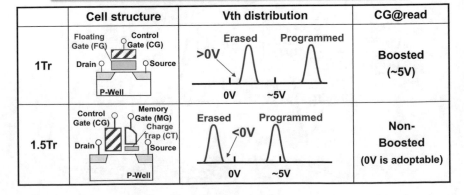

<< Advantages of 1.5Tr (Split-gate) Structure >>
1. **Source-side injection (SSI) programming**
   → **Low-power and fast program**
2. **Deplete-free (no leakage) in unselected WLs**
   → **Low-power and fast read thanks to non-boosted WL**

	Cell structure	Vth distribution	CG@read
1Tr	Floating Gate (FG), Control Gate (CG), Drain, Source, P-Well	Erased / Programmed, >0V, 0V, ~5V	**Boosted (~5V)**
1.5Tr	Control Gate (CG), Memory Gate (MG), Charge Trap (CT), Drain, Source, P-Well	Erased / Programmed, <0V, 0V, ~5V	**Non-Boosted (0V is adoptable)**

Fig. 3.4 Advantages of a 1.5Tr (split-gate) cell

bit-line levels are 0 V in unselected state). In other words, 1.5Tr and 2Tr cell structures can be immune to read disturb if the V_{th} window and the bias applied to each node in cells are properly designed.

3.2.2 Floating-Gate (FG) Flash-Memory Technology

Throughout the history of the semiconductor-manufacturing industry, many LSI chip vendors and foundries have accumulated much knowledge about FG-type cells in terms of process, device, design, and testing method. As a result, FG-type cells have a huge amount of actual use in cases of both stand-alone and embedded flash memories.

Table 3.1 lists the variation of FG-type memory cells categorized into three types according to N_{tr}. A 1Tr FG cell is a simple NOR-type cell suitable for high density where a stacked-gate transistor serves as both a selection gate and data storage. In contrast, in a 1.5Tr FG cell (split-gate FG cell) [2] and 2Tr FG cell [3], two transistors are connected in series where one serves as a selector and the other serves as data storage. They need a bit larger area than the 1Tr FG cell, but they exceed the 1Tr FG cell in performance and power.

Taking a 1Tr cell as an example, Fig. 3.5 illustrates I–V characteristics and channel formation in two different charge stored conditions—programmed cell and erased cell—in read operation. Data stored in a cell are read out by applying an appropriate voltage to the CG. A programmed cell has a higher threshold voltage (V_{th}) due to electrons inside the FG and turns off (OFF state). In contrast, because an erased cell has no or fewer electrons inside the FG than a programmed cell, its V_{th} is lower. Accordingly, an erased cell turns on and flows a certain amount of cell current (ON state).

Table 3.1 Variation of eFlash-memory cells: floating-gate type

Cell structure	1Tr cell (NOR type)	1.5Tr cell [2] (SuperFlash™)	2Tr cell [3]
Program	CHE	SSI	FN
Erase	FN (poly-sub)	FN (poly-poly)	FN (poly-sub)
Device structure			
Advantage	High density	Fast program, Low power read	Low power P/E

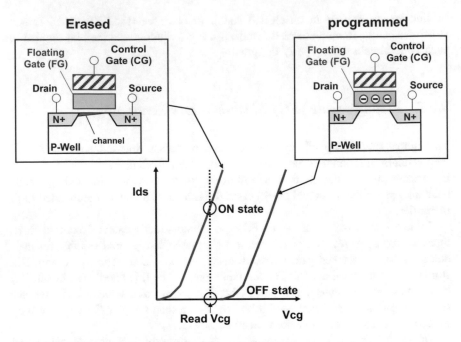

Fig. 3.5 Read-operation mechanism for FG-type cells

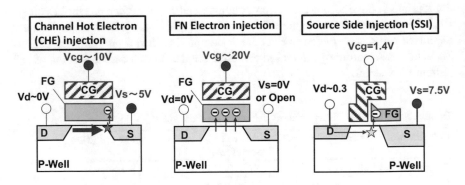

Fig. 3.6 Program-operation principle of FG-type cells

Program-operation mechanism of FG-type cells

In program and erase operations, electrons are injected into or ejected from the FG through an insulator material. Various types of cell structures and physical phenomena for program and erase operations have been proposed and used.

Figure 3.6 summarizes the major electron-injection mechanisms for flash memories.

(1) Chanel hot electron (CHE) injection
By applying higher voltages on the CG and the source, electrons are accelerated at the source side by a horizontal high-electric field and injected into the FG by a vertical high-electric field. The program current is relatively greater than other mechanism.

(2) Fowler–Nordheim (FN) tunneling electron injection
High CG voltage induces a high-electric field at the bottom oxide insulator, thus resulting in electron injection to the FG by Fowler–Nordheim (FN) tunneling phenomena. Compared with CHE injection, the program current is quite small, but a higher voltage is required.

(3) Source-side injection (SSI)
In 1.5Tr cells, high source voltage coupling to the FG and the lower CG voltage (slightly above the threshold voltage) induces a higher electric field at the gap region between the FG and the CG. Because both horizontal and vertical electric fields have maximum field strength at the gap region, hot electrons are generated there with high efficiency and then injected into the FG. Therefore, SSI can achieve smaller program current without degrading programming speed.

Erase-operation mechanism of FG-type cells

In general, there are two different concepts for erase operation: ejecting electrons from the FG or injecting holes into the FG so as to neutralize electrons inside the FG. However, because holes passing through the bottom insulator may cause defects inside the insulator and significantly degrade data-retention performance, basically hole injection is not used in FG-type cells. In contrast, CT-type cells, which are more robust against insulator defects, assertively use hole injection for erase operations as will be discussed later.

Figure 3.7 shows the major FN electron-ejection mechanisms in FG-type cells.

(1) Channel FN electron ejection
Negative high voltage applied to the CG and/or positive high voltage applied to the P-Well induce FN tunneling current flow from the FG to the channel. The

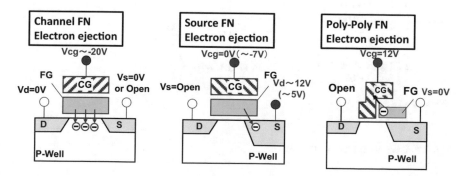

Fig. 3.7 Erase-operation principle of FG-type cells

erase current is quite small, but high voltages are needed. In addition, the well in the memory-array areas should be divided (at least at the size of an erase unit) and biased independently when positive high voltage is used. This leads to a considerable increase in area.

(2) Source FN-electron ejection

Positive high voltage applied to the source line also enables the ejection of electrons from the FG to the source by FN-tunneling phenomena. To apply higher voltage to the source requires a deep source junction to reduce junction leakage, which limits cell scalability.

(3) Poly-to-Poly FN-electron ejection

Electron ejection from the FG poly to CG poly by FN tunneling can be performed by applying positive high voltage on the CG. An electric-field enhancement effect with a needle-shaped structure at the FG corner can reduce FN erase voltage. In addition, no negative voltage is needed in the erase operation.

Program-/erase-operation issues in FG-type cells

An over-erased state (or depleted state) is defined as that threshold voltage of some erased cells that eventually becomes negative by largely crossing over the target-erase level in the erase operation as shown in Fig. 3.8. In the case of a 1Tr cell, over-erased cells can flow a considerable current between the BL and the SL even when they are unselected. As a result, when over-erased cells exist on the selected bit line (BL), both cell current from a selected cell and leakage current from over-erased cells add onto the BL, thus causing read failure.

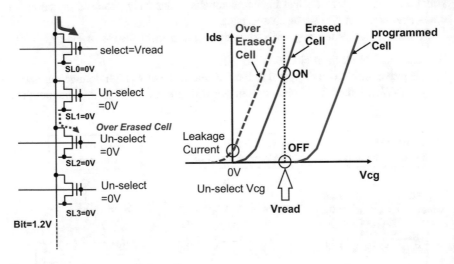

Fig. 3.8 Over-erase issue in 1Tr cells

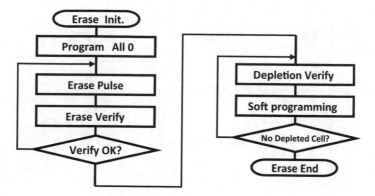

Fig. 3.9 Erase-flow diagram with soft programming

To prevent read failure due to an over-erase issue, complex erase flow should be adopted. Figure 3.9 shows an example of erase flow to avoid over-erase issue with the following steps;

- Program all 0 (high V_{th}): During this step, the cells with low V_{th} in the erase unit are programmed so that the I–V characteristics of all cells in the unit is set as similar as possible before erase operation.
- Erase pulse: All of the cells in the erase unit are biased at predetermined high voltages to erase them for a proper time.
- Erase verify: Verification of the cell state, a sort of read operation, is performed with proper voltage biased on the CG to check whether the V_{th} of all cells becomes lower than the target level. If all of the cells pass, the next step (depletion verify) begins. In contrast, if at least one cell dose not pass, the erase-pulse step starts again. Erase pulse can be repeatedly performed until (1) all of the cells pass erase verify or (2) the pulse count reaches predetermined maximum value.
- Depletion verify: Due to the variation in cell characteristics (e.g., V_{th} distribution, erase speed), some of the cells in the unit may have extremely low V_{th} (<0 V) corresponding to an over-erased state. In this step, the V_{th} of each of the cells is measured to detect over-erased cells.
- Soft programming: Detected deplete cells are slightly re-programmed so as to increase their threshold >0 V.

In contrast to the 1Tr cell, a 1.5Tr cell has an advantage in over-erase issues as shown in Fig. 3.10. Even if the V_{th} of the FG portion becomes <0 V, CG can cut the leakage path to BL by setting properly the V_{th} of the CG. Therefore, a 1.5Tr cell is free from over-erase issues and a V_{th} of the CG <0 V is adoptable. The same holds for a 2Tr cell.

Program disturb is another issue commonly seen in any eFlash cell during program operations. As illustrated in Fig. 3.11, when high voltages are applied to selected cells during program operations, they are also applied to unselected cells

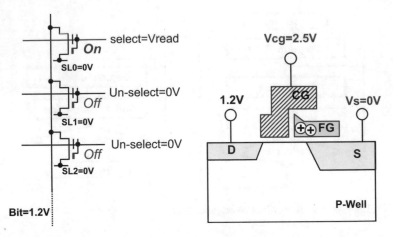

Fig. 3.10 Over-erase free-cell structure

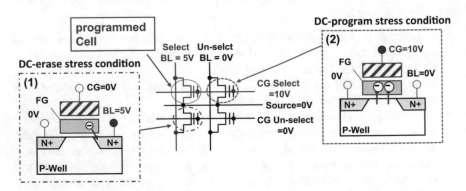

Fig. 3.11 Program-disturb issue

that share nodes (CG or BL). In Fig. 3.11, there are two program-disturb modes to which unselected cells can be subject. The first program-disturb mode works as a weak erase operation with the BLs of cells biased at high voltage and can affect the V_{th} of programmed cells. The second program-disturb mode works as a weak program operation with CGs biased at high voltage and can gradually shift the V_{th} of erased cells.

The influence of program disturb on the V_{th} of unselected cells is determined not only by the voltages used for program operations and the program pulse-duration time to complete program operations but also by memory-array architecture. Proper array division in the consideration of program disturb can minimize the influence as will be discussed later.

Reliability issues in FG-type cells
The Reliability of FG-type cells gradually becomes degraded as program-and-erase operations are repeatedly performed due to degradation of the bottom tunnel oxide.

As for program/erase endurance, it is well-known that the V_{th} difference between programmed cells and erased cells becomes narrower after a number of program/erase (P/E) cycling under the condition that constant P/E voltages and pulse time are used (without V_{th} verification). In Fig. 3.12, an example of V_{th} shift versus number of P/E cycles is depicted. This phenomena is explained by the generation and accumulation of fixed charges in the bottom tunnel oxide due to charge injection and ejection through it. These fixed charges deteriorate the programming efficiency and erase speed, and consequently lead to endurance degradation. Note that this degradation in endurance can be seen even in typical cells.

Another critical reliability issue is the degradation of data retention (more specifically, data loss or V_{th} lowering in programmed cells) along with P/E cycling, which is shown in Fig. 3.13a. This is also due to degradation of the bottom tunnel oxide. Note that these data losses do not occur in typical cells but rather in a small number of singular cells.

High-electric field stress on thin tunnel oxide increases the tunneling (leakage) current under a low-electric field. This low-electric field tunneling current is definitely deviated from FN tunneling current and is called Stress Induced Leakage Current (SILC). SILC is related to the defects in bottom tunneling oxide, which are induced by charge injection under high-field stress. These defects increase traps in the bottom oxide and assist with the tunneling current.

The behavior of gate current in an FG-type cell before and after high-electric field stress is illustrated in Fig. 3.13b. At high-electric field, gate current after high-electric field stress decreases, causing a longer erase time after P/E cycling as explained in Fig. 3.12. On the contrary, at low-electric field the opposite phenomena occurs. Gate current after high-electric field stress increases due to stress-induced leakage current (SILC). Although the degradation of data retention due to SILC is only seen in a small number of singular cells, SILC strongly limits the scaling of the bottom oxide thickness and, as a result, the scalability of FG-type cells.

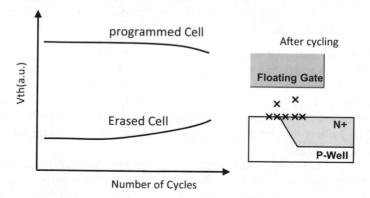

Fig. 3.12 Memory-cell intrinsic endurance characteristics and schematic representation of the state of oxide defect and interface generated by CHE injection

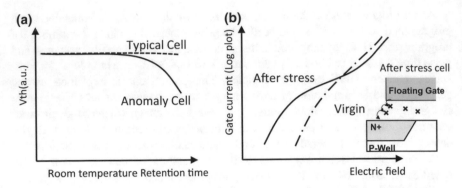

Fig. 3.13 **a** Single-bit data loss after program/erase cycling and **b** stress-induced leakage current (SILC)

1Tr. NOR

Operation	Vd	Vcg	Vs	Vsub
Program	0	PHV	PHV	0
Erase	0	NHV	0	0
Read	Vdd	Vdd	0	0

PHV: Positive High Voltage
NHV: Negative High Voltage
Vdd: Logic power supply or little modified voltage

Fig. 3.14 2Tr FG-type cell and its operation voltages

To suppress the P/E speed variation among cells under various process-voltage–temperature conditions and to relax electrical stress on memory cells, incremental step pulse program and erase (ISPP/ISPE) have been widely used in eFlash memories (not only with FG-type cell but also with other types of cells) as well as in stand-alone flash memories [4]. The details of these techniques and their effects will be explained in Chap. 7.

1Tr-NOR cell, split-gate cell (1.5Tr-cell), and 2Tr-cell

1Tr NOR cell structure and basic operation voltages are shown in Fig. 3.14 [5, 6]. The cell is programmed by CHE injection and erased by channel FN ejection. Due to CHE injection, the program current is larger with more power consumption. A 1Tr cell has the advantage for high density, but a complex erase-flow diagram is needed to prevent over-erase issues. The 1Tr NOR cell will be discussed more extensively in Chap. 4.

A 1.5Tr cell has an advantage in over-erase-free operation. A fast and low-power program operation is achievable thanks to SSI injection, which is investigated in [7] for the first time. A needle-shaped FG induces a high-electric field near the needle portion for poly-to-poly FN-erase operation [2]. This needle-shaped FG enables to

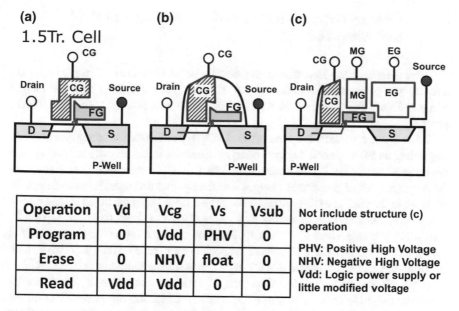

(a)
1.5Tr. Cell

(b)

(c)

Operation	Vd	Vcg	Vs	Vsub
Program	0	Vdd	PHV	0
Erase	0	NHV	float	0
Read	Vdd	Vdd	0	0

Not include structure (c) operation

PHV: Positive High Voltage
NHV: Negative High Voltage
Vdd: Logic power supply or little modified voltage

Fig. 3.15 A 1.5Tr FG-type cell and its operation voltages

use a relatively thick tunnel oxide and improves reliability by suppressing single-bit data loss due to SILC after P/E cycling. However, additional process steps are needed compared to a 1Tr cell due to the CG and the complicated FG shape formation. Figure 3.15 shows the cell-structure transition, along with cell scaling, in one of the most widely adopted 1.5Tr FG cells [2, 8, 9]. A control-gate transistor can be composed by a low-voltage transistor with thin-gate oxide because no high voltages are applied in each operation. Therefore, a read path can be built by low-voltage, high-performance logic transistors, thus resulting in faster read operations than for a 1Tr cell [10]. Details will be presented in Chap. 5.

Regarding the 2Tr cell [2, 11–14], an example of cell structure with an operation-voltage table is shown in Fig. 3.16. Both the program and erase operations use an FN-tunneling mechanism. Compared with the 1Tr cell, the 2Tr cell has larger sizes but this can reduce the overhead of timing and peripheral area associated with the erase flow to avoid over-erase issues (Fig. 3.9) in the same manner as a 1.5Tr cell.

2Tr. Cell

Operation	Vd	Vsg	Vcg	Vs	Vsub
Program	0	0	PHV	0	0
Erase	float	0	NHV	float	0
Read	Vdd	Vdd	Vdd	0	0

PHV: Positive High Voltage
NHV: Negative High Voltage
Vdd: Logic power supply or little modified voltage

Fig. 3.16 A 2Tr FG-type cell and its operation voltages

3.2.3 Charge-Trapping (CT) Flash Technology: SONOS and Nano-Dot

CT-type cells that use thin silicon nitride film or Si nano dots as storage area are another type of typical eFlash cells along with FG-type cells. They have advantages compared with FG-type cells in terms of reliability and affinity with logic CMOS process.

Table 3.2 compares an FG-type cell and a CT-type cell from structural viewpoints. Regarding an FG-type cell, injected charges distribute uniformly in the floating gate which is a conductor. If one conductive defect exists in the oxide film around the floating gate, all of the stored charges are finally lost through this one defect. In contrast, in the case of a CT-type cell, charges are stored by traps in the silicon nitride film and at the interface of nitride and oxide. Because these traps are separately located, only charges near the defect are lost; however, the rest of the charges are not lost. This means that a CT-type cell is intrinsically more reliable than an FG-type cell. In addition, because the height of a CT-type cell is substantially lower than that of an FG-type cell, thanks to thin film storage, a CT-type cell can be integrated within the same gate height as logic transistors even in advanced-technology nodes. This means that a CT-type cell has much better scalability than an FG-type cell from the viewpoints of reliability and affinity to advanced logic CMOS process.

Table 3.2 Advantages of a charge trapping-type cell

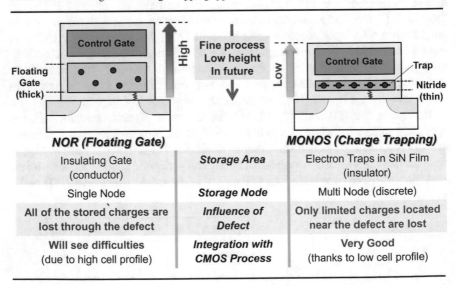

NOR (Floating Gate)	Storage Area	MONOS (Charge Trapping)
Insulating Gate (conductor)	Storage Area	Electron Traps in SiN Film (insulator)
Single Node	Storage Node	Multi Node (discrete)
All of the stored charges are lost through the defect	Influence of Defect	Only limited charges located near the defect are lost
Will see difficulties (due to high cell profile)	Integration with CMOS Process	Very Good (thanks to low cell profile)

Table 3.3   Variation of eFlash-memory cells: charge-trapping type

Cell structure	1Tr cell [15] (SONOS)	1.5Tr cell [16] (SG-MONOS)	2Tr cell [17] (pMOS)
Program	FN	SSI	CHE
Erase	FN	HH	FN
Device structure			
Advantage	High density	Fast program, Low power read	Low power P/E

Table 3.3 lists the variation in CT type cells. In these memory cells, a nitride film or nano-crystals are used for data storage instead of a poly-silicon floating gate. They are also categorized into three types according to the number of Trs (N_{tr}) as is the case of FG-type cells.

The program-operation mechanism of CT-type cells is almost the same as that for FG-type cells. In erase operation, there is one significant difference between both types. Erased FG-type cells store no or fewer electrons in the FG, whereas erased CT-type cells store holes in traps as will be explained in detail later.

Program-operation mechanism of CT-type cells

The program-operation mechanism of a CT-type cell is similar to that of an FG-type cell as FN or channel hot-electron injection. The distribution of injected electrons in the storage layer is quite different between a CT-type cell and an FG-type cell because the charge-trap layer (nitride) in a CT-type cell is an insulator, whereas the FG in an FG-type cell is a conductor. In a CT-type cell, electrons are trapped locally near the injection point and not distributed so broadly as in an FG-type cell. In the case of using channel FN-electron injection, electrons are injected and trapped over the entire channel area. In contrast, when channel hot-electron injection is adopted, electrons are not trapped near the drain (low-electric filed region); instead they are trapped near the source (high-electric field region). In this manner, electron locations in the charge-trap layer strongly depends on the program mechanism. The same holds for the location of holes injected in erase operations in the charge-trap layer. Distribution mismatch between electrons and holes due to their locality of trapped positions should be considered in memory-cell design and operation.

Erase-operation mechanism of CT-type cells

In contrast to the erase operations in FG-type cells, CT type cells assertively adopt hole injection thanks to their robustness against defects in insulator. Typical erase mechanisms used in CT type cells are as follows (see Fig. 3.17).

(1) Channel electron ejection and hole injection

Negative bias applied at CG or positive bias applied at well induce electron ejection and hole injection over the whole channel area. Erase current is quite small, but relatively high voltages are required. In initial erase period, trapped electrons are ejected from trap layer followed by holes being injected from P-Well to trap layer for the later period of erase. CT type cell become lower V_t than neutral V_t due to holes trapped after erase operation.

Figure 3.17 shows schematic energy-band diagrams at erase condition. Electron ejection can occur by the following mechanisms: (1) Trapped electrons are emitted to the conduction band by FN tunneling through the bottom oxide; and (2) FN tunneling from the SiN-trap site to Si-sub through the SiN and the bottom oxide-potential barrier. On the contrary, holes are injected by direct-tunneling mechanism because a CT-type cell uses ultra-thin bottom oxide (approximately 2 nm) compared with the FG-type thicker bottom oxide (approximately 9 nm).

(2) Band-to-band tunneling (BTBT) hot-hole injection

The combination of positive high voltage on the source and negative high voltage on the MG generates hole-electron pairs by BTBT. Compared with channel-hole injection, BTBT hot-hole injection achieves faster erase operation with lower voltages. On the contrary, a larger erase current per cell is consumed. In addition, the hole-injection area is limited near the source junction in this mechanism. To compensate for this locality of hole distribution, CHE injection or SSI program mechanism is often used in combination with BTBT hot-hole injection for erase.

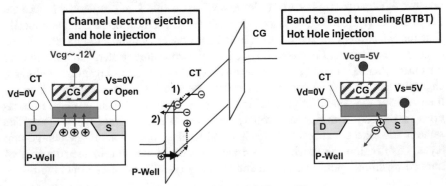

1) Electron emission to conduction band and through bottom oxide
2) Electron FN tunneling from SiN trap site to Si-sub

Fig. 3.17 Erase-operation principle of CT-type cells

Program/erase operation issues in CT-type cells

One of the key points in program and erase operations of CT-type cells is how to match both distributions of electrons and holes considering the locality of their trapped positions in the charge-trap film. A large distribution mismatch between electrons and holes induces excess stress during program/erase operations and causes degradation of data retention after large number of program/erase cycles due to the charge redistribution of electrons and holes. Figure 3.18 illustrates the concept of charge mismatch in a CT-type cell [18].

In case of using hot-carrier operations, such as CHE or SSI electron injection and BTBT hole injection, a mismatch issue of charge distribution can occur because electrons and holes are injected with a slightly different distribution in the horizontal direction. Accelerated electrons through the channel are injected into the charge-trap film with relatively spread distribution at the high-electric field near the junction edge. In contrast, the BTBT mechanism is initiated just inside the source n+ junction under a high-electric field between negative gate voltage and positive junction voltage. As a result, holes are locally injected above the source n+ junction edge as shown in Fig. 3.18. This mismatch can be controlled and reduced by appropriate bias conditions of the CG and source during program and erase operations [19]. Furthermore, so-called step-pulse biasing works effectively to minimize the distribution mismatch between electrons and holes.

Reliability of CT-type cells

In an FG-type cell, retention issues result from anomaly cell leakage induced by SILC after cycling. The leakage finally drains all of the charges in a cell because the FG is a conductive material. On the contrary, charges stored in a CT-type cell are locally trapped inside a nitride film or at the interface of nitride and oxide. As a result, only those near the defect are lost; however, the rest of the charges are not lost. Therefore, a CT-type cell is intrinsically reliable in terms of data retention. Thanks to high reliability, the CT-type cell not only demonstrates high scalability, but it also can use relatively thinner bottom tunneling oxide without regard to data loss, thus leading to another benefit of decreasing operation voltages in program and erase operations.

Fig. 3.18 Mismatch of
CT-type cell charge

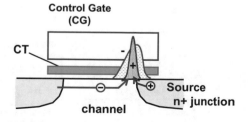

In general, a relatively large V_{th} shift is observed at the early stage of retention in the majority of CT-type cells. This phenomenon can be well attributed to the following three factors: (1) electron/hole recombination generated by charge mismatches; (2) recovery of the interface state (silicon/bottom oxide) generated by the program/erase operations; and (3) charge detrap from shallow traps [20]. After this stage, the V_{th} decay rate becomes sufficiently small.

Brief overview of CT-type cells

A 1Tr SONOS (poly-silicon oxide–nitride–oxide–silicon) cell in Fig. 3.19 is a simple charge trap type memory cell using ONO film under the control gate instead of floating gate [15, 21]. This simple structure needs less additional mask and process steps compared with an FG-type cell. Because FG-type cell needs extra process steps to isolate each FG and to control effective voltage coupling with CG. P/E operations of 1Tr SONOS cell are based on FN tunneling mechanism. Ultra-thin bottom oxide, approximately 2 nm, for direct tunneling is adapted to use FN-electron injection and hole injection. Although thinner-bottom oxide raises concerns about reliability degradation in FG-type cells, it is applicable even to CT-type cells for automotive applications thanks to the defect immunity of charge-trap property.

As is the case in 1Tr FG-type cells, the 1Tr SONOS cell also has the same issue of over-erase as shown in Fig. 3.8. Furthermore, due to ultra-thin bottom oxide, those using positive CG bias in read operations suffer from read disturb, which is equivalent to "weak program" and causes a V_{th} shift of erased cells after larger read cycles. Read disturb generates great concern in case of automotive applications where higher reliability is strongly required. Design countermeasures by dedicated array architecture will be discussed in Chap. 6.

The NROM® pictured in Fig. 3.20 is a unique CT-type cell that employs two separate physical charge packets in order to realize a 2-bit/cell storage [18, 19, 22]. NROM technology is suitable for data, code, and embedded flash applications with low-cost fabrication process. The charge-storage area consists of three stacked layers, bottom tunneling oxide, nitride film, and top oxide (ONO). Program and erase mechanisms are channel hot-electron (CHE) injection and band-to band tunneling (BTBT) hot-hole injection, respectively. Charges are stored at the narrow regions near the source and drain junctions.

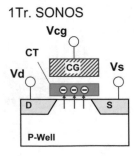

Operation	Vd	Vcg	Vs	Vsub
Program	NMV	PMV	NMV	NMV
Erase	PMV	NMV	PMV	PMV
Read	0	0	Vdd	NV

PMV: Positive Middle Voltage
NMV: Negative Middle Voltage
Vdd: Logic power supply or little modified voltage
NV: Negative Low Voltage

Fig. 3.19 A 1Tr SONOS cell and its operation voltages

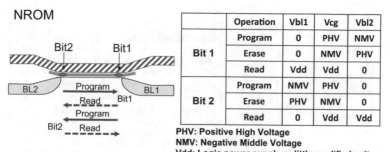

	Operation	Vbl1	Vcg	Vbl2
	Program	0	PHV	NMV
Bit 1	Erase	0	NMV	PHV
	Read	Vdd	Vdd	0
	Program	NMV	PHV	0
Bit 2	Erase	PHV	NMV	0
	Read	0	Vdd	Vdd

PHV: Positive High Voltage
NMV: Negative Middle Voltage
Vdd: Logic power supply or little modified voltage

Fig. 3.20 NROM cell and its operation voltages

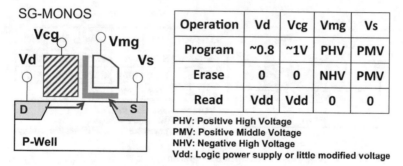

Operation	Vd	Vcg	Vmg	Vs
Program	~0.8	~1V	PHV	PMV
Erase	0	0	NHV	PMV
Read	Vdd	Vdd	0	0

PHV: Positive High Voltage
PMV: Positive Middle Voltage
NHV: Negative High Voltage
Vdd: Logic power supply or little modified voltage

Fig. 3.21 SG-MONOS cell and its operation voltages

Read operation is performed by interchanging the roles of source and drain in a cell, which is called as "reverse read." Thanks to charge storage within a narrow region, a readout of 1 bit on the source side is not affected by the other bit on the drain side under proper drain-source voltage (V_{ds}).

A 1.5Tr cell, split-gate MONOS (SG-MONOS), architecture is also widely used in CT-type cells shown in Fig. 3.21. Source-side injection (SSI) and band-to-band tunneling (BTBT) hot-hole injection are adopted to inject a charge into the ONO-trap layer in program and erase operations, respectively [16, 23–25]. Details of SG-MONOS technology will be explained in Chap. 7.

Another approach for the CT-type cell has been developed using nanocrystals or nano-dots instead of floating gate or nitride film. Charges are stored by confinement within nanocrystals or nano-dots [26, 27]. Figure 3.22 shows a cross-sectional image of split-gate thin-film storage (SG-TFS) cell in combination with split-gate structure. Here, storage layer is composed by silicon nanocrystals, which are deposited as a mono-layer two-dimensional array and sandwiched between the bottom tunneling oxide and top oxide. The energy barrier height to the bottom and top oxide and the coulomb blockade phenomenon between neighboring nanocrystals stably confine the charges in each nanocrystal. Note that the spacing

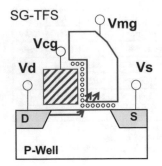

Operation	Vd	Vcg	Vmg	Vs
Program	2uA	0.8	PHV	PMV
Erase	0	0	PHV	0
Read	Vdd	Vdd	Vdd	0

PHV: Positive High Voltage
PMV: Positive Middle Voltage
Vdd: Logic power supply or little modified voltage

Fig. 3.22 An Si-nanocrystal split-gate memory cell and its operation voltages

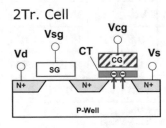

Operation	Vd	Vsg	Vmg	Vs	Vsub
Program	NMV	0	PMV	NMV	NMV
Erase	PMV	PMV	NMV	PMV	PMV
Read	Vdd	Vdd	0	0	0

PMV: Positive Middle Voltage
NMV: Negative Middle Voltage
Vdd: Logic power supply or little modified voltage

Fig. 3.23 A 2Tr SONOS cell and its operation voltages

between nanocrystals should be set beyond the critical distance (approximately 3 nm) to prevent direct tunneling among them.

Thanks to spatially localized charge storage inside each nano-crystal, an SG-TFS cell retains its overall data state even if one of the nanocrystals fails to store charges. This unique feature provides from defect immunity as well as charge-trap memory cell.

In the operation voltage table in Fig. 3.22, source-side injection is used for programming and FN tunneling erase while positive voltage applied to the MG minimizes oxide degradation. This scheme is advantageous by eliminating any negative high voltage in all operations. Read operation is performed in the same way as SG-MONOS except $V_{mg} = V_{dd}$.

An example of a 2Tr SONOS cell structure with operation voltage table is shown in Fig. 3.23 [17, 28]. Both of program and erase operations use an FN-tunneling mechanism. Compared with a 1Tr cell, a 2Tr cell has larger sizes, but it can reduce the overhead of timing and peripheral area associated with the erase flow to avoid over-erase issue (Fig. 3.9) as well as the 1.5Tr cell.

3.2.4   Other Miscellaneous Embedded Flash Technologies: EEPROM, CMOS-Flash, and One-Time Programmable (OTP) Cell

In this section, embedded programmable ROM cell technologies other than Flash in Fig. 3.2 are briefly overviewed. They have their own features such as high endurance (EEPROM) or compatibility with logic CMOS process (OTP, MTP). The memory cells of OTP and MTP are based on logic CMOS transistors and therefore can be implemented in logic CMOS process with no cost adder. OTP and MTP IPs are often available at the same time as the launch of logic CMOS technology. Their main applications include (1) identification or encryption data storage for security functions and (2) parameter storage for timing adjustment or function selection. In addition, EEPROM is used for data logging in the field.

EEPROM Technology

As an electrically erasable programmable read-only memory (EEPROM), FLOating-gate Thin Oxide (FLOTOX) is the world's first memory cell that adopts an FN-tunneling mechanism for both program and erase operations [29]. Figure 3.24 shows a cross-sectional image of a FLOTOX cell including a select Tr. In program operation, high voltage is applied on the control gate and select gate, and source and drain voltages are set at 0 V. Electrons are injected from the drain to the FG through partially thin oxide (8 to approximately 10 nm) on the drain-junction region. On the contrary, in erase operation, SG and drain are biased at high voltage, which ejects electrons under floating-gate thin oxide.

OTP technology

One-time programmable (OTP) ROM has been widely used in various products to store such information as trimming parameters or product codes. Aside from the programmability after the completion of the wafer process, another advantage of OTP is that it is composed of logic CMOS transistors and can be implemented with no additional mask and process step to logic CMOS process.

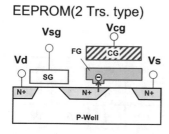

EEPROM(2 Trs. type)

Operation	Vd	Vcg	Vsg	Vs	Vsub
Program	0	PHV	PHV	0	0
Erase	0	0	PHV	PHV	0
Read	0	Vdd	Vdd	Vdd	0

PHV: Positive High Voltage
Vdd: Logic power supply or little modified voltage

Fig. 3.24 EEPROM cell structure

A CMOS anti-fuse cell, one of the major OTP memory cells, consists of two n-MOSFETs as shown in Fig. 3.25, program transistor (anti-fuse line [AF]), and select transistor (SG). Cell state is defined in response to the resistance from the AF to the source in a selected cell. "Anti-fuse" means that programmed cells show low resistance, which is opposite to general eFuse (metal fuse) whose resistance becomes high after programming. Bias conditions in program and read operations for OTP cell are described in Fig. 3.25.

In program operation, high voltage applied to the AF induces gate oxide breakdown at the edge of the program transistor, thus resulting in low resistance in the path from the AF to the source. In read operation with Vread applied to the AF, cell current from the AF to the source is detected or compared with the reference current.

Another OTP memory cell is the FG-type cell as illustrated in Fig. 3.26. It is composed of two p-MOSFETs connected in series, a floating-gate transistor on right side, and a select transistor on the left side [30, 31].

In program operation, an FG-type OTP memory cell uses a channel hot hole to inject hot electrons into the gate of the floating-gate transistor (FG). A lateral high-electric field accelerates holes and induces electron–hole pair generation. Finally, some of the electrons are injected into the FG by a coupling effect between the FG and the drain. Electrons stored in the FG turn on the floating-gate transistor.

Another OTP cell shown in Fig. 3.27 uses a CHE-injection mechanism for programming. Some of the electrons accelerated by the drain high-electric field are injected into a gate insulator and side-wall spacer, which increase threshold voltage

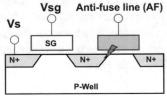

Operation	Vs	Vsg	AF	Vsub
Program	0	Vdd	PHV	0
Read	0	Vdd	Vdd	0

PHV: Positive High Voltage
Vdd: Logic power supply or little modified voltage

Fig. 3.25 A 2Tr anti-fuse cell

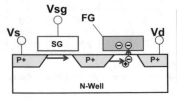

Operation	Vs	Vsg	Vd	Vsub
Program	PHV	0	0	PHV
Read	Vdd	0	0	Vdd

PHV: Positive High Voltage
Vdd: Logic power supply or little modified voltage

Fig. 3.26 A 2Tr FG-type cell

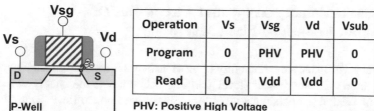

Operation	Vs	Vsg	Vd	Vsub
Program	0	PHV	PHV	0
Read	0	Vdd	Vdd	0

PHV: Positive High Voltage
Vdd: Logic power supply or little modified voltage

Fig. 3.27 A charge trapping-type cell

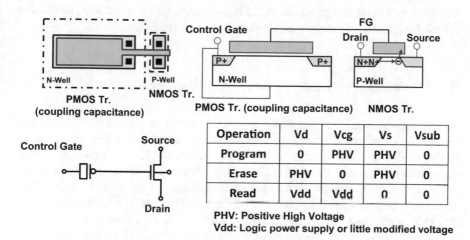

Operation	Vd	Vcg	Vs	Vsub
Program	0	PHV	PHV	0
Erase	PHV	0	PHV	0
Read	Vdd	Vdd	0	0

PHV: Positive High Voltage
Vdd: Logic power supply or little modified voltage

Fig. 3.28 Schematics of a logic-NVM multi-time program erase cell structure

and, as a result, decrease the drain current of the cell. These electrons are stably trapped there even at high temperature [32].

MTP technology

Figure 3.28 shows the schematic and cross-sectional image of a multi-time-programmable (MTP) erase cell that can be implemented in standard logic CMOS process without additional process steps [33, 34]. The cell consists of a PMOS transistor as coupling capacitor and a NMOS transistor as readout path. Both transistors share a poly gate, which acts as floating gate. A connected PMOS well and P+ diffusion node applied positive bias works the control gate. The single-poly flash cell is programmed by CHE injection applied V_{PP} at the drain and the control gate. In contrast, the erase operation uses FN tunneling applied under high voltage at the drain and the source.

3.3 Basic Memory Array and Circuit Design of an Embedded Flash Macro

The total performance and reliability of an eFlash system is determined by not only the potential of memory cells (level 1 in Fig. 3.1) but also the quality of memory array and related peripheral circuits in the eFlash hard macro (level 2) used in the system. Reasonable choice of design concept, methodologies, and techniques for memory array and related peripheral circuits is important to obtain the maximum performance from memory cells.

In this section, basic memory-array architecture is briefly overviewed at first followed by circuit-design topics of (1) high-voltage (HV) circuit design, (2) read circuit design, and (3) design techniques for higher reliability.

3.3.1 Basics of Memory—Array Architecture

Figure 3.29 illustrates a simplified block diagram of memory-array and related circuits in an eFlash hard macro. Memory array is divided into multiple blocks based on the specification of erase unit size. Block (BLK) #0 to #M are normal erase units to store user data. Extra BLK #0 to #N are dedicated to special purposes such as boot-code storage, system-parameter storage, or redundancy. Each block has corresponding row (X) decoders and drivers. They are activated according to the address information to access a certain portion of memory array and selectively apply high voltages to corresponding nodes in the selected block during program and erase operations. Note that all of the cells in a block are simultaneously erased. In contrast, in general, a column (Y) path, including BLs, is shared by multiple blocks. Figure 3.29 shows that BLs, column selector (Y-sel), column-control circuits composed by data latch (DL), and sense amplifier (SA) are shared by all of the blocks. In some cases, especially to target higher read performance, hierarchical BL and/or SA architecture is adopted to reduce parasitic resistance and capacitance of BLs and minimize data-propagation time to output buffers. The number of DL and SA is basically the same as maximum program-unit size and read-bus width, respectively. Y-sel has a predetermined multiplex ratio and connects each DL or SA to a certain BL according to the column address.

Taking 1Tr FG cells as an example, Fig. 3.30 shows a typical array architecture in an erase unit. Here, each CG has its dedicated driver and is connected to the gate of multiple cells. In contrast, the SL is connected to the source of all of the cells in the block. In addition, in this case because the well of selected cells should be biased at high voltage, the well of memory array is also divided for each block. (There is no need of well division if the well bias is fixed.) As already mentioned, BLs are shared by not only the cells in this block but also by those in other blocks.

The basic concept of memory array architecture for eFlash is similar to that of stand-alone flash if the same cell type is adopted in both cases. However, in some

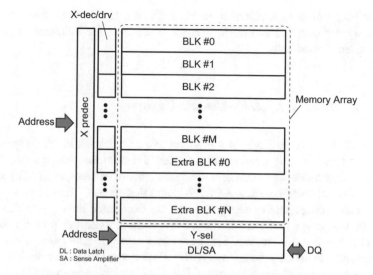

Fig. 3.29 A simplified block diagram of flash array and related circuits

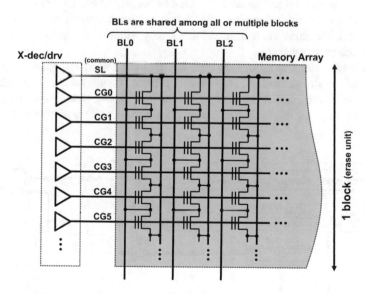

Fig. 3.30 A typical array architecture in an erase unit with 1Tr FG cells

cases—such as automotive or industry applications—higher performance and re-liability are strongly required. More CG or SL stich regions may be necessary for faster rise/fall time or noise suppression. Regarding reliability, sharing nodes among cells causes program disturb. In general, the more cells share one node to which high voltage is applied in P/E operations, the more vulnerable the cells in the block to program disturb, resulting in V_{th} shift and, in the worst case, data loss. Finer array division and control is needed to suppress the influence of program disturb to unselected cells.

3.3.2 High Voltage (HV)-Circuit Design

Figure 3.31 shows an example of a high-voltage (HV) circuit block diagram in an eFlash hard macro. The power unit includes the reference voltage and current generator, ring oscillator (ROSC), and several sets of charge-pump (CP) subsystems, regulators, voltage switches (VSWs), and dischargers. The output signals of the reference voltage and current generator are distributed not only to the circuit blocks in the power unit but also to other blocks outside the power unit (e.g., sense-amplifier control circuits). The base clock signal is generated in the ROSC and distributed to the CP subsystems. Each CP subsystem includes a dedicated local clock generator, a level detector, and a CP to generate a predetermined HV level. In some cases, a regulator is connected to the output of a CP subsystem and down-converts its level to another level. One of the VSWs whose outputs are shared (wired-ORed) selectively turns on and transfers the corresponding voltage level to the decoders. The decoders and the drivers finally supply those voltages to selected nodes in the memory array.

HV circuit blocks in an eFlash hard macro look almost the same as those in stand-alone flash memories. However, different circuit-design techniques or methodologies are needed for some circuits in a power unit. These differences partly come from the constraints in the specifications of HV transistors available for

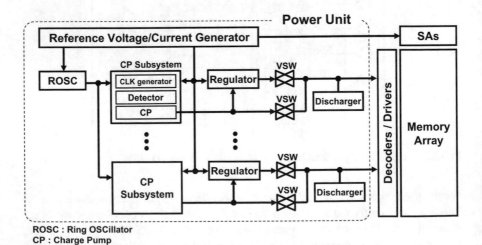

ROSC : Ring OSCillator
CP : Charge Pump
VSW : Voltage SWitch

Fig. 3.31 A simplified block diagram of high-voltage circuits

HV-circuit design. In stand-alone flash memories are developed dedicated HV transistors, whose breakdown voltages (V_{bd}) are designed to be sufficiently higher than the maximum voltages necessary for program or erase operations. In contrast, in the case of embedded flash memories, the integration of such dedicated HV transistors means a significant increase of total process cost. Although it is desirable to use standard IO transistors already equipped in base logic CMOS process, their maximum breakdown voltages are not high enough for HV-circuit design in flash memories. This is why peculiar design techniques are needed in eFlash hard-macro design to balance the voltage requirements from the memory cell operations and the specifications of existing devices.

Here are described three examples of HV circuit-design techniques in eFlash hard macros. Figure 3.32 presents the first basic technique to handle high voltage. When handling high voltage, VP = VHH, which is higher than V_{bd}, low-side voltage VN is boosted from 0 V to VLH so that VP − VN (=VHH − VLH) becomes smaller than V_{bd}. Note that adequate timing control is needed so that VN reaches VLH earlier than or concurrently with the time VP reaches the VHH from the VHL.

Figure 3.33 shows the level-shift circuit used in a VSW as the second example. Protection transistors (P3, P4, N3, and N4) are inserted with voltage-mitigation signals VMP and VMN biased to their gate. N3 and N4 clamp the voltage level of nodes NA and NB so that their maximum value is less than VMN − V_{thn}. P3 and P4 clamp the voltage level of VOUTP and NC so that their maximum value is higher than VMP + V_{thp}. The voltage difference between arbitrary two nodes is one of the following values; 0, VP−(VMN − V_{thn}), VMN − V_{thn}, VP−(VMP + V_{thp}), or VMP + V_{thp}. By properly choosing the voltage of VMN and VMP, these values can be set less than V_{bd}. Similar design techniques have been reported in technical conferences and widely adopted especially in embedded memories [35, 36].

CP circuits also need similar kind of design techniques under the breakdown voltage-constraint of MOS capacitors, whose gate-oxide thickness is the same as that in existing IO transistors in the base logic CMOS process. An example of such design techniques is shown in Fig. 3.34. Serially connected MOS capacitors are used at higher-voltage stages to mitigate the bias to each capacitor ($<V_{bd}$). Aside

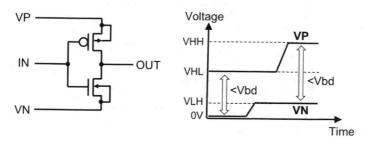

Fig. 3.32 HV-design techniques (1)

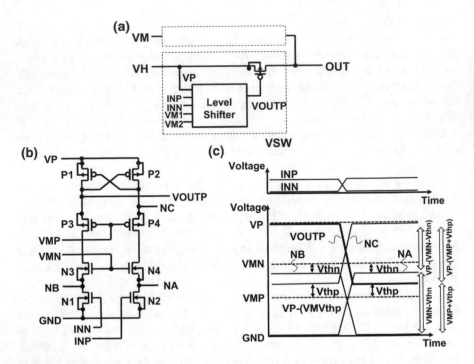

Fig. 3.33 HV-design techniques (2): **a** block diagram of VSW, **b** circuit example of level shifter in VSW, and **c** operation-timing chart

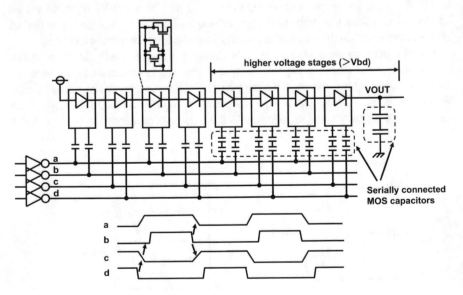

Fig. 3.34 HV-design techniques (3)

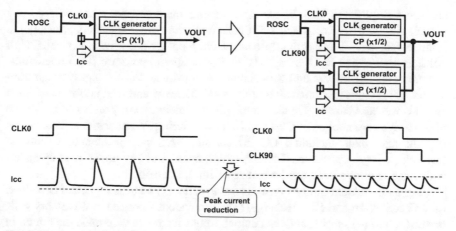

Fig. 3.35 An example of a low-noise charge-pump structure

from this example, metal-oxide-metal (MOM) capacitors with widely spaced metals are also applicable without additional process cost.

Note that these design techniques have a disadvantage of area penalty over the case where the dedicated devices with higher breakdown voltage are introduced. Because the introduction of additional devices into base logic CMOS process definitely increases total process cost, designers should make a final judgment regarding high-voltage circuit-design policy between two options: (1) use standard IO devices in base logic CMOS process with no additional process cost but larger area size; or (2) introduce dedicated HV devices with decreased area size but additional process cost.

Other constraints may pose a challenge to HV-circuit design in eFlash hard macros. For example, when noise-sensitive analog IPs, such as analog-digital converter (ADC), are implemented with eFlash hard macros on the same die, some measures to suppress power-line noise caused by CP operations in eFlash hard macros should be carefully investigated. One of the effective design measures is to divide one CP into two half-size CPs and operate them in different phases to each other as shown in Fig. 3.35. In this scheme, total output current from CPs per one cycle of the CP clock is not changed, but the peak of current consumption from power lines (Icc) can be reduced by half. Therefore, this scheme can reduce the noise on power lines and ensure the stable and accurate operations of noise-sensitive IPs even when eFlash macros operate in parallel with those IPs on the same die.

3.3.3   Read-Circuit Design

The memory cell structure and operation principle strongly influence the choice of peripheral transistors to build memory array and related circuits such as decoders and sense amplifiers (SAs). Figure 3.36 presents the comparison of memory-array organization with 1Tr cells and 1.5Tr cells (the case with 2Tr cells is basically the same as that with 1.5Tr cells). In case of the memory array with 1Tr cells, high voltages are applied to CGs and BLs during program and erase (P/E) operations. Accordingly, CG drivers and SAs should be constructed using high-voltage transistors. This results in degradation of the area efficiency and read performance such as random access latency. On the contrary, in the memory array with 1.5Tr cells, no high voltages are applied to CGs and BLs even during P/E operations. This means that the read path, including CG drivers and SAs, can be composed only of low-voltage logic transistors. Therefore, high speed and low power read operation with small area overhead can be realized with 1.5Tr cells.

In general, readout timing consists of four parts: (1) address input and decoding, (2) CG activation and BL precharge, (3) data readout from selected memory cells onto BLs, and (4) sensing and data output. Major factors to determine the timing in each part are array division (more specifically, the number of memory cells per CG or BL) and sensing operation. As for array division, the smaller the number of memory cells per CG or BL is set, the shorter the time of (2) or (3) can be. However, finer division of the memory array leads to greater area overhead.

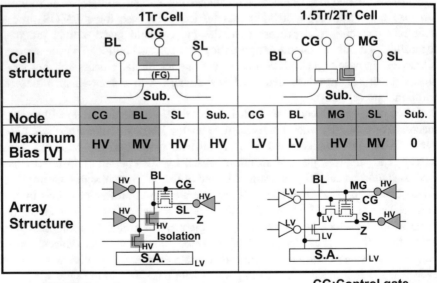

Fig. 3.36 Array structure comparison between 1Tr and 1.5Tr cells

The balance of read performance and area overhead should be taken into account in the actual macro design.

Sensitive read schemes, including SAs with small offset, are indispensable because correct data can be read out even with a smaller cell current difference between erased cells and programmed cells after long retention time. In other words, narrower V_{th} window between the program state and the erase state is acceptable with sensitive read schemes, thus resulting in less electrical stress to memory cells in P/E operations. The necessary cell-current difference is partly determined by the read margin of SAs and the temperature dependency mismatch between the cell current and the reference current.

In Fig. 3.37, three circuit techniques to reduce the necessary cell current difference are presented. The first one is to decrease the necessary read margin with offset cancellation or offset-tolerant operation. The second one is to use the reference current, which compensates for the voltage and temperature dependency of the cell current. One of the most popular ways based on this concept is to use dummy cells to generate the reference current [37]. The third one is a complementary cell scheme, in which a programmed cell and an erased cell compose one logical bit, and each cell serves as reference to the other. Therefore, a very narrow V_{th} window between the program state and the erase state is acceptable with this scheme.

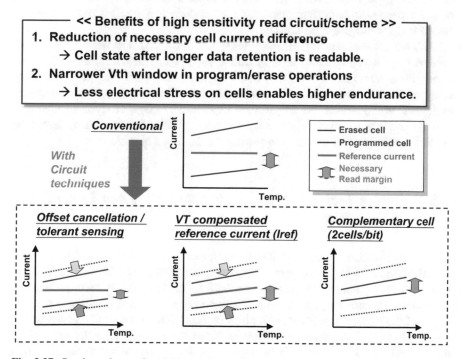

Fig. 3.37 Sensing schemes for highly precise read-out operation

Fig. 3.38 An example of an offset-tolerant sense amplifier (in current sampling phase) [35]

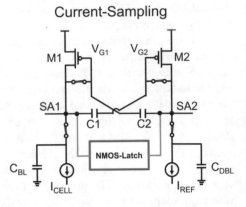

Figure 3.38 shows an example of a sense amplifier with offset-tolerant operation [38]. Two cross-coupled pairs of a pMOS transistor and a capacitor are used for both current sampling and current-ratio amplifying. In the current-sampling phase, M1 and M2 operate in a diode-connected manner. Their gate-source voltages correspond to cell current (I_{cell}) and reference current (I_{ref}), respectively, and are stored in C1 and C2. Thanks to this architecture, M1 and M2 can flow the same current as I_{cell} and I_{ref} despite V_{th} and/or BL-capacitance mismatch at the beginning of the current-ratio amplifying phase. As a result, a very small current difference between I_{cell} and I_{ref} can be sensed.

Although those sensing schemes shown in Fig. 3.37 are quite effective to achieve high accuracy in sensing operation even with smaller current difference, they present some disadvantages. First, they need additional area in the memory array or the SA region. Second, offset cancellation in an analog manner requires additional timing before the actual amplification of voltage or current difference, and this causes relatively slow access latency. These disadvantages should be studied carefully so as to meet the priority in the target specification.

3.3.4 Design Techniques for Higher Reliability and Lower Failure Rate

In some applications (e.g., automotive), extremely high reliability and low failure rate after shipment (e.g., sub-ppm failure) are strongly required. To meet with these requirements, additional design efforts should be made in eFlash hard macros.

Program disturb, V_{th} shift in unselected memory cells during program operation, is one of the unavoidable issues in flash memories because one or two HV nodes in an eFlash-memory cell share the drivers with many other cells. Here, let us consider the case in Fig. 3.39, where 2Kbit memory cells are connected to the common source line and are sequentially selected for programming in four divided manner (512 bits × 4 cycles). High voltage is supplied to the common source line in program operations. Assuming that the typical program time is T_{pa}, the program-disturb time for unprogrammed cells is $4*T_{pa}$ on average. This example implies that program-disturb time becomes longer when the number of cells sharing a node is set larger or the number of cells selected at the same time is set smaller. This problem can be solved or alleviated by physically dividing the node into multiple units and activating each unit in a sequential manner as shown in the bottom part of Fig. 3.39.

Proper DFM (design for manufacturing) and DFT (design for testing) should be applied in eFlash hard-macro design to secure higher yield and screen potential failures that could emerge in the field after shipment. Figure 3.40 shows some examples of DFM and DFT suitable for eFlash hard macro to achieve a very low failure rate. Redundant via is widely adopted to reduce the open-via failures. Regarding test functions, conventional static or dynamic burn-in (screening) tests have been widely used in non-volatile memories. These tests use special high voltages and multi-block-selection functions to shorten the test time. As a new trend of eFlash testing, SCAN test has been gradually introduced to efficiently detect failures in the random logics in eFlash hard macros.

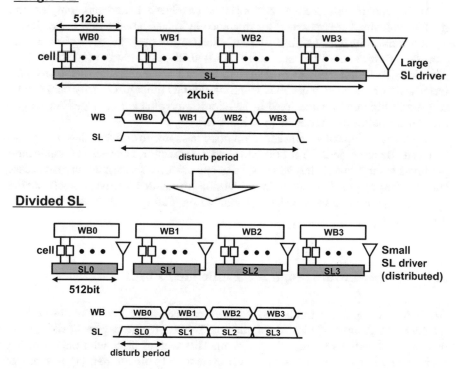

Fig. 3.39 A disturb-conscious array architecture

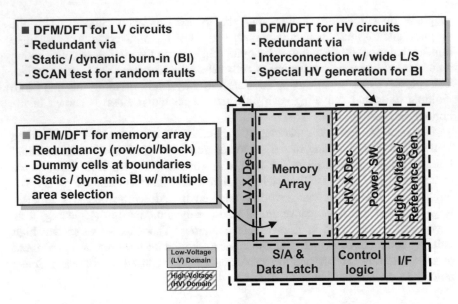

Fig. 3.40 A DFM/DFT for low failure rate in an eFlash hard-macro design

At the end of this section, examples of an eFlash hard-macro pin list and specification sheet are presented for reference in Tables 3.4 and 3.5, respectively.

Compared with the interface of stand-alone memories, the constraints on the number of pins are not so stringent. Rather, as shown in Table 3.4, a non-multiplexed address and parallel data bus are commonly used for better performance. For example, a data-bus width of 256 bits (plus ECC parity bits) can be adopted in high-end applications. In addition, some pins are equipped for simple and direct control of various internal functions.

In Table 3.5, typical specifications for code-flash macro and data-flash macro are included. Because code flash is used to store instruction code and/or application programs at large size, it has large capacity and achieves faster random read access and P/E throughput. On the contrary, data flash provides relatively small rewrite size and higher endurance, which are suitable for storing temporary or frequently updated data.

3.4 System-Level Design for Embedded Flash

As shown in Fig. 3.1, an eFlash sub-system consists of one or more eFlash hard macros and a group of dedicated soft macros. These soft macros give flexibility and diversity of functions to the eFlash system. This section deals with the following topics related to the functions of eFlash system and soft macros; (1) low power design (intermittent operation), (2) reliability enhancement on a system level, (3) safety function including ECC, and (4) security function.

Table 3.4 Example of an eFlash hard-macro pin list

Category	Pin name	Direction	Description
Bus	XAD[x − 1:0]	in	X (row) address input (x bits)
	YAD[y − 1:0]	in	Y (column) address input (y bits)
	D_{IN}[d − 1:0]	in	Data input (d bits)
	D_{OUT}[d − 1:0]	out	Data output (d bits)
	CLK	in	Clock signal
	WEN	in	Bus write enable
	EMATEN	in	Extra mat access enable
	REGEN	in	Register access enable
Power	V_{dd}	in	Power supply for core transistors
	V_{CC}	in	Power supply for I/O (HV) transistors
	V_{REF}	in	Reference voltage
	V_{SS}	in	Ground
System control	RESET	in	Reset signal
	STBY	in	Stand-by mode transition signal
P/E control	PRG	in	Activation of program sequence
	ERS	in	Activation of erase sequence
	RZB	out	P/E sequence status (ready/busy)
Test mode	TMDEN	in	Test mode enable
	SCAN	in	Scan test enable
	VMON	In/out	Internal voltage monitor/external force

Table 3.5 Example of an eFlash hard macro-specification sheet

Item		Spec.	
		Code	Data
Operation temperature (junction)		−40–125 °C (Consumer) −40–150 °C (Automotive)	
Capacity per macro		~2 MB	~128 KB
Read	Bus width	128 bits	32 bits
	Random access freq.	20–100 MHz	~10 MHz
Program	Program unit size	8B–256B	4B–128B
	Program time (typ.)	~2 s / 1 MB	~200 μs / 8B
Erase	Erase unit size	8KB–256KB	1KB–16KB
	Erase time (typ.)	~5 s / 1MB	~200 ms / 16KB
Endurance		1K–10K cycles	125K cycles
Retention[a]		10 years (Consumer) 20 years (Automotive)	

[a]After 1K cycles in code / 125K cycles in data. Temperature profile should be taken into consideration

3.4.1 Low-Power Design: Intermittent Operation

To realize a greener society, low power consumption is becoming essential even in high-end applications. One of the major design countermeasures for low power consumption is intermittent operation with a power-gating scheme. In a power-gating scheme, circuits in a chip are divided into several power domains, and the power in each domain can be controlled independently from others with power switches to connect its own local power line to a global power line. Here are two examples: SRAM-based operation and eFlash-based operation (Fig. 3.41).

In SRAM-based operation, there occurs a large standby current—especially under high temperature—due to large leakage current because SRAM is a volatile memory, and power must always be supplied to SRAMs. On the contrary, in eFlash-based operation, the standby current is much smaller because the power supply to the domain including eFlash can be fully cut off thanks to its non-volatility. In this context, eFlash-based operation is more suitable to achieve low power consumption with intermittent operations. However, in eFlash-based operation, it takes much time for start-up as well as data evacuation because of internal voltage generation including high voltages for P/E operations at the beginning of active period and the long programming time to eFlash at the end of active period. These timing overheads may cause large additional power consumption during the active period. This implies that there will be "break-even

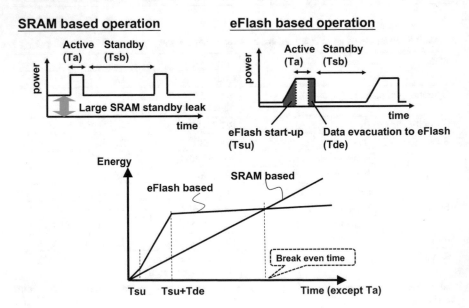

Fig. 3.41 Intermittent operation with SRAM or flash

point" in selecting eFlash-based intermittent operation to achieve lower power consumption as expected. In this respect, an eFlash system with shorter wake-up time and shorter programming time is more desirable. In addition, read-only intermittent operation with no need of high voltage generation will expand the application area of eFlash-based intermittent operations.

3.4.2 Reliability Enhancement on a System Level

An eFlash sub-system can enhance data reliability as a system beyond the original memory cell's characteristics. Two examples are shown in Fig. 3.42.

The first example uses the dedicated SRAM buffer, whose size is N_S times larger than the user-programming size [39]. Data from users are temporary stored in the SRAM buffer until it is filled. Then all of the data in the SRAM buffer are programmed to a certain area in the eFlash-memory array at the same time. The next time the SRAM buffer is filled, the data in the buffer are programmed to another area in the eFlash-memory array. The dedicated eFlash soft macro administrates the destination area for the next data transfer from the SRAM buffer. Assuming that the ratio of the eFlash-memory array size and the SRAM-buffer size is N_F, and the equivalent endurance from a user viewpoint can be enhanced by a factor of N_S multiplied by N_F compared with that of memory cells.

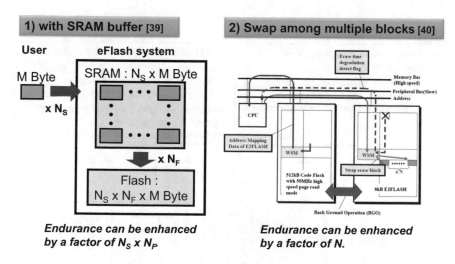

Fig. 3.42 An eFlash system for higher data reliability [39, 40]

In the second example [40], multiple erase blocks in the eFlash-memory array are used to swap an erase block to others, by which users can enjoy equivalently higher endurance in a limited area. The erase time of a block is monitored. When it becomes larger than a certain value, the flag bit in the block is activated. If the CPU detects the activation before the start of the next erase operation, it swaps the block to the new one.

3.4.3 Safety Function Including ECC

Safety and security requirements have been becoming remarkable not only in automotive applications but also in consumer/IoT (Internet Of Things) applications. To implement sufficient safety and security functions, hard macros and soft macros in an eFlash sub-system should cooperate properly as described in Fig. 3.43.

Examples of safety functions required to eFlash systems are summarized in Fig. 3.44.

ECC (error correction code) is the most widely used safety method not only in the stand-alone NVMs such as NAND but also in the embedded NVMs. Table 3.6 lists the features of ECC implementation in stand-alone flash memories and embedded flash memories for high-end MCU applications.

Because eFlash has smaller page size, the ratio of parity bits to data bits becomes relatively large, resulting in large area penalty. Another feature is the speed requirement on ECC decoding. In general, one of the reasons to implement eFlash

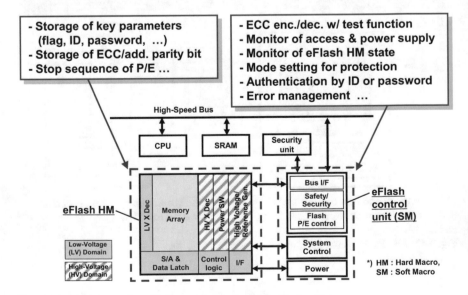

Fig. 3.43 The role-sharing of safety and security in an eFlash system

■ **Error detection / correction (w/ redundant bits)**
  - **Data parity bits for ECC**
  - **Address parity bits for decoder fault detection**
  - **Test function for ECC decode path**
     e.g. Fault detection by "error injection"

■ **Data protection related to P/E operation** *) P/E = Program/Erase
  - **Safe update of key parameters or boot code**
  - **Block/area lock setting against unintended P/E**
  - **Monitor of illegal command issue**
  - **Protection against power failures during P/E**

Fig. 3.44 Examples of safety function in an eFlash system

Table 3.6 Implementation of the ECC function

		eFlash for high-end	Stand-alone flash (NAND)
Area	Read unit or page size	Small ($\sim$16B)	Large ($\sim$1 KB)
	Ratio of parity bits to data bits	Large (>10%)	Small (2–5%)
Speed	Random access time	Short (<10 ns)	Long ($\sim$50 µs)
	Timing budget for ECC decode	Small (2–3 ns)	Large ($\sim$10 µs)

on the same die with CPUs is to achieve greater performance. Especially in automotive MCU applications, faster code and/or data fetch from eFlash is critical for real-time operations. Therefore, considering area and speed, simple ECC with fewer parity bits and shorter decoding time is preferable for eFlash. In other words, intrinsic reliability of memory cells has much more importance in eFlash.

Two cases of ECC decode timing are shown in Fig. 3.45. In case of one-cycle access, the data output delay in eFlash hard macros and the following ECC decoding time must be within the same one cycle. In case of $N + 1$-cycle access, which is often the case at high frequency, ECC decoding time must be within one cycle next to the array-access cycles. In high-end uses, the timing budget for ECC decoding is no more than several nano-seconds. Therefore, the data output path of eFlash, including ECC decoding, requires one of the most critical timing designs in the entire chip.

ECC can be also used to enhance reliability by correcting the information from cells whose actual reliability is marginal or does not meet the criteria. However, implementation of ECC for enhancement of safety and reliability costs more area and access latency.

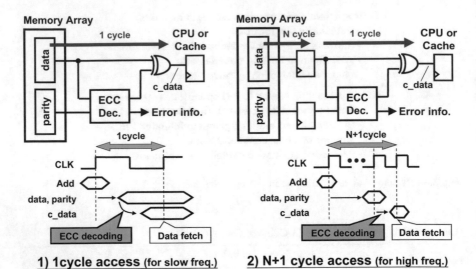

Fig. 3.45 Access path including ECC decode

3.4.4 Security Function

Figure 3.46 shows the security trend of eNVMs in MCU applications. The first major advancement occurred when the embedded memory was changed from mask ROM to eFlash. The contents in eFlash can be updated at garage. To protect the contents against malicious attacks, several functions have been introduced in the eFlash system based on each vendors' security policy. Now, the second major advancement is coming. Because the contents in eFlash can be updated in real time by way of vehicle-to-vehicle or vehicle-to-infrastructure communications, securer protections than before at the whole chip level are strongly required. In addition, common security policies are being discussed in projects such as EVITA.

To prevent illegal data fetch and/or revision in an eFlash system, such security functions as authentication by ID or password and data protection, in collaboration with the chip-level security unit, are implemented in the latest eFlash system as summarized in Fig. 3.47.

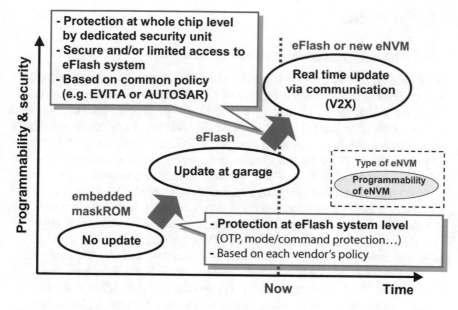

Fig. 3.46 The security trend of eNVM in MCU applications

To prevent important program and data from being illegally fetched or revised, the following security functions are needed;

✓ **Protection against attacks via e-Flash programming environment**
✓ **Protection against attacks to SW and data for security unit**

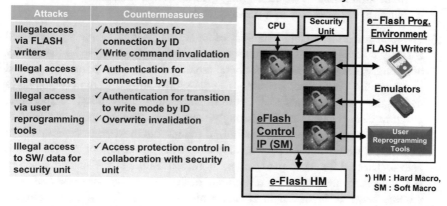

Attacks	Countermeasures
Illegal access via FLASH writers	✓ Authentication for connection by ID ✓ Write command invalidation
Illegal access via emulators	✓ Authentication for connection by ID
Illegal access via user reprogramming tools	✓ Authentication for transition to write mode by ID ✓ Overwrite invalidation
Illegal access to SW/ data for security unit	✓ Access protection control in collaboration with security unit

Fig. 3.47 Examples of a security solution in an eFlash system

3.5   Conclusion

To end chapter 3, the following conclusions are drawn:

(1) In general, eNVM systems should be carefully and hierarchically designed for each level of memory-cell design (level 1), memory array and related peripheral circuit design (level 2), and system-level design in combination with hard macros and dedicated soft macros (level 3). Although the memory cell largely determines the basic electrical and reliability characteristics of the system, proper choice of peripheral circuits can greatly enhance total electrical and reliability characteristics of the system. In addition, toward the upcoming Internet-of-Everything (IoE) era, adequate combination of hard macros and soft macros in the system is essential not only to achieve higher performance and reliability but also to meet wide range of customer requirements regarding secure data storage and access in an eNVM system.

(2) In the history of semiconductor LSI chip research and development, many different kinds of memory cells for programmable ROM have been proposed, commercialized, and manufactured. This diversity of memory cells is still more apparent in embedded uses than in stand-alone uses because (1) those memory cells should be integrated into the existing base process; and (2) requirements from applications are also diversified and, especially, reliability and performance take higher priority in embedded uses compared with stand-alone uses. The memory cells developed for embedded uses in logic CMOS process are approximately classified according to three viewpoints:

• Necessity of different transistors from existing logic CMOS transistors
• Structure of storage layer: floating-gate or charge-trapping
• Number of transistors in a memory cell: 1, 1.5, or 2

(3) The total performance and reliability of an eFlash system is largely determined not only by the potential of the memory cells themselves but also by the quality of memory array and related peripheral circuits in the eFlash hard macros used in the system. As for memory-array design, proper array division is essential to meet the requirements of performance and reliability. Especially, program-disturb characteristics strongly depend on memory-array architecture. Regarding peripheral circuit design, different circuit-design techniques or methodologies are needed for some circuits in the power unit of an eFlash hard macro. These differences partly come from constraints in the specifications of HV transistors available for HV-circuit design. From the viewpoint of total process cost, it is desirable to substitute standard IO transistors already equipped in base logic CMOS process for such HV transistors, but their maximum breakdown voltages are not high enough for HV-circuit design in flash memories. Aside from HV-circuit design, the read-access path design and DFT/DFM also requires careful choice of circuit configuration and design methodologies.

(4) Proper combination of eFlash hard macros and soft macros in an eFlash system greatly enriches its characteristics and functionality. Low power consumption by intermittent operations and equivalently greater endurance by block management are good examples of characteristic enhancement by a sort of hardware–software co-design on an eFlash-system level. Security and safety functions are and will be strongly required in the upcoming IoE era for secure data storage and access from outside.

The rest of this book is organized as follows. Chapter 4 describes the history and technologies of 1Tr NOR-type eFlash memory including discussions from the viewpoints of some specific applications. Chapter 5 provides the details of split-gate FG-type eFlash memory from various perspectives. Chapter 6 is dedicated to 1Tr CT-type eFlash memory with the latest topics including its automotive application. Finally, Chap. 7 presents split-gate CT-type eFlash memory and its advantages in detail.

Notes on terminology and notations
Throughout this book the following notations rules are applied.

(1) In memory capacity, "bit" are often abbreviated as "b" and "Byte" as "B."
(2) SONOS is consistently used for poly-Si–oxide–nitride–oxide–silicon cell structure, often denoted as MONOS with the same meaning in the literature.
(3) The split-gate cell is often denoted as a 1.5-Tr cell because this is an intermediate structure between 1Tr and 2Tr cells.
(4) The terms "source" and "drain" are defined as the node from which carriers flow out and the node through which carriers leave, respectively, based on semiconductor physics. In the read operation of an n-channel memory, electrons flow from the source to the drain with 0 V biased at source and positive voltage biased at drain (BL). However, during program operation, drain (BL) is biased at 0 V or V_{dd} (dependent on program data), and source is biased at higher voltage to induce CHE or SSI electron injection. Thus, a physical node serves as source in one case and a drain in other case. Traditionally, the node notation of source and drain based on read-current flow has been widely accepted. Therefore, this notation style is used in this book, except for a few explanations about program mechanism in Chap. 4.

References

1. K. Baker, Embedded nonvolatile memory technology. in *Proceedings of IEEE International Conference on IC Design & Technology*, pp. 185–189 (2009)
2. S. Kianian, A. Levi, D. Lee, Y.-W. Hu, A novel 3 volts-only, small sector erase, high density flash E2PROM. in *Symposium on VLSI Technology, Digest of Technical papers*, pp. 71–72 (1994)
3. Y.K. Lee, J.H. Moon, Y.H. Kim, M.-J. Chun, S.-Y. Ha, S. Choi, H. Yoo, H. Jeon, J. Yu, J.-U. Han, E. Jung, C. Chung, 2T-FN eNVM with 90 nm logic process for smart card. in

Non-Volatile Semiconductor Memory Workshop and International Conference on Memory Technology and Design, pp. 26–27 (2008)

4. K.-D. Suh, B.-H. Suh, Y.-H. Lim, J.-K. Kim, Y.-J. Choi, Y.-N. Koh, S.-S. Lee, S.-C. Kwon, B.-S. Choi, J.-S. Yum, J.-H. Choi, J.-R. Kim, H.-K. Lim, A 3.3 V 32 Mb NAND flash memory with incremental step pulse programming scheme. in *IEEE Solid-State Circuits Conference, Digest of Technical Papers*, pp. 128–129 (1995)

5. F. Piazza, C. Boccaccio, S. Bruyere, R. Cea, B. Clark, N. Degors, C. Collins, A. Gandolfo, A. Gilardini, E. Gomiero, P. M. Mans, G. Mastracchio, D. Pacelli, N. Planes, J. Simon, M. Weybright, A. Maurelli, High performance flash memory for 65 nm embedded automotive application. in *IEEE International Memory Workshop*, pp. 77–79 (2010)

6. H. Kojima, T. Ema, T. Anezaki, J. Ariyoshi, H. Ogawa, K. Yoshizawa, S. Mehta, S. Fong, S. Logie, R. Smoak, D. Rutledge, Embedded flash on 90 nm logic technology & beyond for FPGAs. in *IEEE International Electron Devices Meeting*, pp. 677–680 (2007)

7. M. Kamiya, Y. Kojima, Y. Kato, K. Tanaka, Y. Hayashi, EPROM cell with high gate injection efficiency. in *IEEE International Electron Devices Meeting*, pp. 741–744 (1982)

8. R. Mih, J. Harrington, K. Houlihan, H.K. Lee, K. Chan, J. Johnson, B. Chen, J. Yan, A. Schmidt, C. Gruensfelder, K. Kim, D. Shum, C. Lo, D. Lee, A. Levi, C. Lam, 0.18 μm modular triple self-aligned embedded split-gate flash memory. in *Symposium on VLSI Technology, Digest of Technical papers*, pp. 120–121 (2000)

9. D. Shum, J.R. Power, R. Ullmann, E. Suryaputra, K. Ho, J. Hsiao, C.H. Tan, W. Langheinrich, C. Bukethal, V. Pissors, G. Tempel, M. Röhrich, A. Gratz, A. Iserhagen, E.O. Andersen, S. Paprotta, W. Dickenscheid, R. Strenz, R. Duschl, T. Kern, C.T. Hsieh, C.M. Huang, C.W. Ho, H.H. Kuo, C.W. Hung, Y.T. Lin, L.C. Tran, Highly reliable flash memory with self-aligned split-gate cell embedded into high performance 65 nm CMOS for automotive & smartcard applications. in *IEEE International Memory Workshop*, pp. 139–142 (2012)

10. L.Q. Luo, Y.T. Chow, X.S. Cai, F. Zhang, Z.Q. Teo, D.X. Wang, K.Y. Lim, B.B. Zhou, J.Q. Liu, A. Yeo, T.L. Chang, Y.J. Kong, C.W. Yap, S. Lup, R. Long, J.B. Tan, D. Shum, N. Do, J.H. Kim, P. Ghazavi, V. Tiwari, Functionality demonstration of a high-density 1.1 V self-aligned split-gate NVM cell embedded into LP 40 nm CMOS for automotive and smart card applications. in *IEEE International Memory Workshop*, pp. 165–168 (2015)

11. G. Tao, S. Nath, Experimental study of temperature dependence of program/erase endurance of embedded flash memories with 2T-FNFN NOR device architecture. in *IEEE-IRW*, pp. 76–79 (2006)

12. A. Conte, G. Matranga, D.D. Costantini, M. Micciche, C. Ucciardello, A.D. Martino, F. Granata, A. Castagna, P. Zuliani, E. Gomiero, R. Annunziata, J. Devin, F. Maugain, J. Acland, J. Sonzogni, A 90 nm embedded page flash for EEPROM replacement in system on chip. in *IEEE Non-Volatile Semiconductor Memory Workshop*, pp. 28–30 (2008)

13. A. Baiano, M. van Duuren, E. van der Vegt, B. Schippers, R. Beurze, D.T. Mofrad, H. van Zwol, Y. Chen, J. Chiang, H. Lokker, K. van Dijk, J. Verbree, Y.N. Chen, J. Garbe, R. Verhaar, D. Dormans, Junction optimization for embedded 40 nm FN/FN flash memory. in *IEEE International Memory Workshop*, pp. 173–176 (2015)

14. S.-R. Kim, K.J. Han, K.-S. Lee, R. Li, J. Wolfman, T.-H. Kim, P. Liu, H. Kim, P.-Y. Lee, Y. Wang, Y. Jia, F. Dhaoui, F. Hawley, H.-C. Tseng, High performance 65 nm 2T-embedded Flash memory for high reliability SOC applications. in *IEEE International Memory Workshop*, pp. 158–160 (2010)

15. S. Minami, Y. Kamigaki, A novel MONOS nonvolatile memory device ensuring 10-year data retention after 10^7 erase/write cycles. IEEE Trans. Electron Devices **40**(11), 2011–2017 (1993)

16. W.M. Chen, C. Swift, D. Roberts, K. Forbes, J. Higman, B. Maiti, W. Paulson, K.T. Chang, A novel flash memory device with split gate source side injection and ONO charge storage stack (SPIN). in *Symposium on VLSI Technology*, pp. 63–64 (1997)

17. H.M. Lee, S.T. Woo, H.M. Chen, R. Shen, C. D. Wang, L.C. Hsia, C.C.-H. Hsu, NeoFlash®—true logic single poly flash memory technology. in *IEEE Non-Volatile Semiconductor Memory Workshop*, pp. 15–16 (2006)
18. M. Janai, B. Eitan, A. Shappir, E. Lusky, I. Bloom, G. Cohen, Data retention reliability model of NROM nonvolatile memory products. IEEE Trans. Device Mater. Reliab. **4**(3), 404–415 (2004)
19. A. Shappir, E. Lusky, G. Cohen, I. Bloom, M. Janai, B. Eitan, The two-bit NROM reliability. IEEE Trans. Device Mater. Reliab. **4**(3), 397–403 (2004)
20. Y. Kawashima, T. Hashimoto, I. Yamakawa, Investigation of the data retention mechanism and modeling for the high reliability embedded split-gate MONOS flash memory. in *IEEE International Reliability Physics Symposium*, MY.6.1–MY.6.5 (2015)
21. H. Mitani, K. Matsubara, H. Yoshida, T. Hashimoto, H. Yamakoshi, S. Abe, T. Kono, Y. Taito, T. Ito, T. Krafuji, K. Noguchi, H. Hidaka, T. Yamauchi, A 90 nm embedded 1T-MONOS flash macro for automotive applications with 0.07 mJ/8kB rewrite energy and endurance over 100 M cycles under Tj of 175 °C. in *IEEE Solid-State Circuits Conference, Digest of Technical Papers*, pp. 140–141 (2016)
22. B. Eitan, P. Pavan, I. Bloom, E. Aloni, A. Frommer, D. Finzi, Can NROM, a 2-bit, trapping storage NVN cell, give a real challenge to floating date cells? in *International Conference on Solid State Devices and Materials*, pp. 522–524 (1999)
23. F. Ito, Y. Kawashima, T. Sakai, Y. Kanamaru, Y. Ishii, M. Mizuno, T. Hashimoto, T. Ishimaru, T. Mine, N. Matsuzaki, H. Kume, T. Tanaka, Y. Shinagawa, T. Toya, K. Okuyama, K. Kuroda, K. Kubota, A novel MNOS technology using gate hole injection in erase operation for embedded nonvolatile memory applications. in *Symposium on VLSI Technology, Digest of Technical papers*, pp. 80–81 (2004)
24. T. Kono, T. Ito, T. Tsuruda, T. Nishiyama, T. Nagasawa, T. Ogawa, Y. Kawashima, H. Hidaka, T. Yamauchi, 40 nm embedded SG-MONOS flash macros for automotive with 160 MHz random access for code and endurance over 10 M cycles for data. in *IEEE Solid-State Circuits Conference, Digest of Technical Papers*, pp. 212–213 (2013)
25. Y. Taito, M. Nakano, H. Okimoto, D. Okada, T. Ito, T. Kono, K. Noguchi, H. Hidaka, T. Yamauchi, A 28 nm embedded SG-MONOS flash macro for automotive achieving 200 MHz read operation and 2.0 MB/s write throughput at Tj of 170 °C. in *IEEE Solid-State Circuits Conference, Digest of Technical Papers*, pp. 132–133 (2015)
26. J.A. Yater, S.T. Kang, R. Steimle, C.M. Hong, B. Winstead, M. Herrick, G. Chindalore, Optimization of 90 nm split gate nanocrystal non-volatile memory. in *IEEE Non-Volatile Semiconductor Memory Workshop*, pp. 77–78 (2007)
27. S. T. Kang, B. Winstead, J. Yater, M. Suhail, G. Zhang, C.-M. Hong, H. Gasquet, D. Kolar, J. Shen, B. Min, K. Loiko, A. Hardell, E. LePore, R. Parks, R. Syzdek, S. Williams, W. Malloch, G. Chindalore, Y. Chen, Y. Shao, L. Huajun, L. Louis, S. Chaw, High performance nanocrystal based embedded flash microcontrollers with exceptional endurance and nanocrystal scaling capability. in *IEEE International Memory Workshop*, pp. 131–134 (2012)
28. K. Ramkumar, I. Kouznetsov, V. Prabhakar, K. Shakeri, X. Yu, Y. Yang, L. Hinh, S. Lee, S. Samanta, H.M. Shih, S. Geha, P.C. Shih, C.C. Huang, H.C. Lee, S.H. Wu, J.H. Gau and Y.K. Sheu, A scalable, low voltage, low cost SONOS memory technology for embedded NVM applications. in *IEEE International Memory Workshop*, pp. 119–202 (2013)
29. E. Harari, L. Schmitz, B. Troutman, S. Wang, A 256-bit nonvolatile static RAM. in *IEEE Solid-State Circuits Conference, Digest of Technical Papers*, pp. 108–109 (1978)
30. C.-Y. Lin, C.-H. Lin, C.-H. Ho, W.-W. Liao, S.-Y. Lee, M.-C. Ho, S.-C. Wang, S.-C. Huang, Y.-T. Lin, C.C.-H. Hsu, Embedded OTP fuse in CMOS logic process. in *IEEE International Workshop on Memory Technology, Design, and Testing*, pp. 13–15 (2005)
31. eMemory NeoBit, http://www.ememory.com.tw/html/products_neobit.php. Nov 2016
32. NSCore, http://www.nscore.com/images/WhitePaper_081002.pdf. Nov 2016
33. K. Ohsaki, N. Asamoto, S. Takagaki, A single poly EEPROM cell structure for use in standard CMOS processes. IEEE J. Solid-State Circ. **29**(3), 311–316 (1994)

34. R.J. McPartland, R. Singh, 1.25 volt, low cost, embedded flash memory for low density applications. in *Symposium on VLSI Circuits, Digest of Technical Papers*, pp. 158–161 (2000)
35. S.-H. Song, K.C. Chun, C.H. Kim, A logic-compatible embedded flash memory featuring a multi-story high voltage switch and a selective refresh scheme. in *Symposium on VLSI Circuits, Digest of Technical Papers*, pp. 130–131 (2012)
36. G. Fredeman, D. Plass, A. Mathews, K. Reyer, T. Knips, T. Miller, E. Gerhard, D. Kannambadi, C. Paone, D. Lee, D. Rainey, M. Sperling, M. Whalen, S. Burns, A 14 nm 1.1 Mb embedded DRAM macro with 1 ns access. in *IEEE Solid-State Circuits Conference, Digest of Technical Papers*, pp. 316–317 (2015)
37. C. Demi, M. Jankowski, C. Thalmaier, A 0.13 μm 2.125 MB 23.5 ns embedded flash with 2 GB/s read throughput for automotive microcontrollers. in *IEEE Solid-State Circuits Conference, Digest of Technical Papers*, pp. 478–617 (2007)
38. M.-F. Chang, S.-J. Shen, C.-C. Liu, C.-W. Wu, Y.-F. Lin, S.-C. Wu, C.-E. Huang, H.-C. Lai, Y.-C. King, C.-J. Lin, H.-J. Liao, Y.-D. Chih, H. Yamauchi, An offset-tolerant current-sampling-based sense amplifier for sub-100 nA-cell-current nonvolatile memory. in *IEEE Solid-State Circuits Conference, Digest of Technical Papers*, pp. 206–208 (2011)
39. An example of this scheme is shown in the following web site: http://cache.nxp.com/files/32bit/doc/app_note/AN4282.pdf?fsrch=1&sr=1&pageNum=1. Mar 2015
40. S. Kawai, A. Hosogane, S. Kuge, T. Abe, K. Hashimoto, T. Oishi, N. Tsuji, K. Sakakibara, K. Noguchi, An 8kB EEPROM-emulation DataFLASH module for automotive MCU. in *IEEE Solid-State Circuits Conference, Digest of Technical Papers*, pp. 508–632 (2008)

Chapter 4
Floating-Gate 1Tr-NOR eFlash Memory

Antonino Conte, Fabio Disegni, Francesco La Rosa
and Alfonso Maurelli

4.1 Technology Introduction

The one-transistor cell has been by far the most popular cell in the world of flash NOR stand-alone memories mainly due to its smallest cell size and its capacity to achieve the best possible cost/performances trade-off for a relatively large amount of memory, which is unique to that kind of market.

Consequently, when the memory array size is relatively large (an absolute value cannot be given because that depends on the technology node [1]), 1Tr-NOR architecture has also been the first and one of the best solutions for embedded NVM applications and has been progressively replacing EEPROM cells. In this chapter, embedded multi-megabit flash implementation will be reviewed.

In particular, among the various possible topics [2], the focus has been on the description of how the features offered by the 1Tr-NOR cell can be efficiently integrated into MCUs, which are one of the most popular product families based on embedded flash capability. Consequently to the selection of that market field, it is

A. Conte
Microcontrollers and Digital ICs Group, STMicroelectronics,
Stradale Primosole 50, 95121 Catania, CT, Italy
e-mail: antonino.conte@st.com

F. Disegni · A. Maurelli (✉)
Automotive and Discrete Group, STMicroelectronics,
Via Olivetti 2, 20864 Agrate Brianza, MB, Italy
e-mail: alfonso.maurelli@st.com

F. Disegni
e-mail: fabio.disegni@st.com

F. La Rosa
Microcontrollers and Digital ICs Group, STMicroelectronics,
190 Avenue Celestin Coq, 13106 Rousset Cedex, France
e-mail: francesco.larosa@st.com

© Springer International Publishing AG 2018
H. Hidaka (ed.), *Embedded Flash Memory for Embedded Systems: Technology,*
Design for Sub-systems, and Innovations, Integrated Circuits and Systems,
DOI 10.1007/978-3-319-55306-1_4

important to spend some time describing the way in which integration of the 1Tr-NOR cell can be achieved: this is particularly relevant because the integration of the cell in a high-performance CMOS (such as that needed by MCUs for demanding applications) has been increasingly challenging (starting, for example, at 0.5 μm with the integration of CMOS-optimized titanium salicide within the 1Tr-NOR flash architecture [3]). Furthermore, with technology-node scaling and increased complexity of the basic CMOS, reusing the huge IP portfolio coming with it has become a must, and—as a consequence—the electrical features of the CMOS must also be preserved when embedding a flash macrocell (the flash macrocell includes not only the flash cell but also all of the high-voltage (HV) elementary components needed to manage flash cell operations).

Although the world of embedded flash has plenty of dedicated solutions, there are relatively few published papers; thus, the risk of being too generic is large if describing theoretical solutions. Therefore, the authors prefer to give a more exact overview of concrete solutions implemented at their end, i.e., at STMicroelectronics.

There is no intention of claiming the superiority of the implementations described herein: no memory cell is clearly superior to any others for whatever application, and nor is there a design implementation more superior whatever the flash-cell architecture adopted. However, the 1Tr-NOR flash-cell solution does have some advantages that justify the investments made by some important microelectronics companies in the past decades and continuing in the future. Among the advantages, most relevant are its simplicity, the relatively low implementation costs, and the very high level of robustness widely demonstrated in retention and reliability.

In contrast, in many applicative cases, those technology advantages bring several challenges in terms of design complexity, because 1Tr-NOR use is not obvious when low-power consumption and high performance are required at the same time.

In this book, different solutions for this kind of flash-cell structure and design architectures are described as well as are the different reasons why those solutions have been implemented, all of them with success and satisfaction of the respective proposing groups.

Embedded 1Tr-NOR flash memories have been on the market since the early 1990s. At that time, the most popular solution for stand-alone memory applications was the 1Tr-NOR for technology nodes of approximately 1 μm, and companies working on stand-alone NVM products as well as MCUs, such as STMicroelectronics, naturally promoted it. In fact, the very fast learning curve about yield and failure mechanisms derived by the large volumes called for by the stand-alone market was a huge advantage.

Actually, mastering both the technology and design in-house at the same time enabled a virtuous cycle that leverages the intimate knowledge of the cell peculiarities with optimized design techniques, something not only relevant in terms of the achieved performances, such as access time and power consumption, but also (and more importantly perhaps) for the attainment of the best-possible quality and reliability targets. That is of particular value when dealing with NVM in general and especially when targeting relatively large memory sizes and ppm quality targets such as those in the automotive field.

Since flash cells were first developed, then, it has been clear that some of the process and electrical parameters, key to the memory cell's functionality and reliability, had to remain almost untouched with respect to stand-alone memory technologies when embedded in the chosen CMOS environment as well as when scaling the technology. That has been both a limiting factor for the scaling and at the same time a relevant advantage in granting reliability to the embedded flash-product reliability, even for products with limited volumes and/or harsh automotive requirements. (From this point of view, it is worth remembering that automotive-grade products with embedded 1Tr-NOR flash had already been introduced in the 1990s [4]).

Such key bricks, such as tunnel oxide and interpoly dielectric, together with "sealing" cell process steps, have been the focus of embedded flash-technology development. This sealing has been even more important when dealing with 1Tr-NOR architecture because of whatever defect or imperfection would have not been supported/masked by a selector transistor.

With the increasingly aggressive scaling of the CMOS as a consequence of the increased performances requested by the products, and with the associated complexity of the related basic IPs (standard cells, SRAM, etc.), the possibility of tuning the CMOS around the needs of the flash memory has progressively disappeared. Companies dealing with embedded flash have had to imagine the integration of the flash-process steps with the CMOS steps in a "modular" fashion (see, as an example, the schematic flow reported in Fig. 4.1 for the implementation proposed by STMicroelectronics for the 1Tr-NOR flash embedded with 180-nm state-of-the-art CMOS [5]).

This process-integration scheme, with further refinements [6], has showed its ability to preserve the unique features of the 1Tr-NOR flash cell as well as to reuse the major IP blocks developed within the pure CMOS-technology platform.

A key role in that process-integration scheme is played by the transistors, which must handle the relatively high voltages (approximately 10 V) necessary for programming and erasing operations. These transistors are often also used to manage 5 V IPs, such as I/Os and analogue circuits, and it is extremely important to minimize the extra costs associated with their implementation. Thus, a way to decouple the HV transistors from the basic CMOS was proposed by STMicroelectronics starting with the 180-nm technology node [7–9]. A typical cross-section of those HV transistors at 180-nm technology node is shown in Fig. 4.2.

With this solution, the polysilicon gate of the HV transistor is also the floating gate of the 1Tr-NOR flash. The HV lightly doped drain (LDD) implants are aligned to the gate, while the source and drain implants (the same for HV and LV transistors) are aligned to the polysilicon layer, completely surrounding the HV transistor gate. That polysilicon layer acts as the gate of the LV CMOS transistors as well as the control gate of the flash cell.

In this way, the features specific to the HV transistors (gate oxide, channel, and LDD implants) can be defined before the LV CMOS transistors are built. Moreover, by playing with the biases of the two polysilicon layers and of the substrate, it is

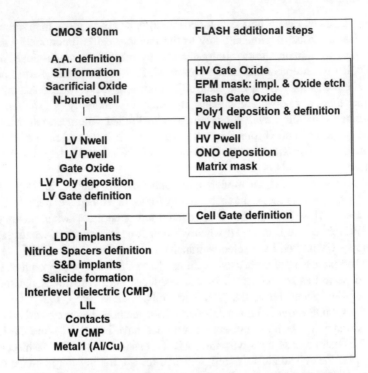

Fig. 4.1 The 180-nm 1Tr-NOR embedded flash schematic process-integration flow

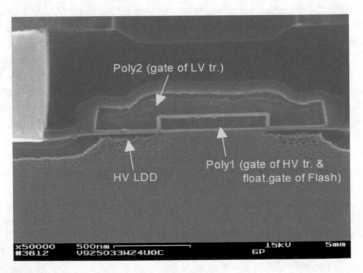

Fig. 4.2 SEM cross-section of an STMicroelectronics HV transistor at 180-nm technology node for managing 5 V IPs as well as flash programming and erasing operations

possible to obtain very effective capacitors, which are especially useful as decoupling capacitors.

The same scheme has also been successfully applied for the technology nodes that followed, with slightly different integration schemes, in order to make the structure more scalable and effective for analogue applications. This is done by removing the polysilicon on top of the HV gate, and consequently the Miller parasitic capacitances, but still keeping the same polysilicon layer that constitutes the floating gate of the flash cell and the gate of the HV transistors.

Meanwhile, as for the progressive technology-node scaling, the 1Tr-NOR flash-cell size has basically escaped the aggressive shrink pursued by the stand-alone memory products. In fact, for stand-alone memories, cell size was largely predominant in determining the product price, thus leading to even more complex process steps and architecture developments (such as the multilevel approach, that is, to store two or more bits/cell) aimed at minimizing the die area while still finding the balance between business advantage and technology extra costs. In contrast, for embedded flash-memory products, at least three factors must be taken into account for the majority of the applications:

- **Overall array size**, which generally represents approximately only 1/5th or 1/6th of the entire chip size; thus, any additional cost put into reducing the flash size is "paid" for by the whole chip.
- **Performance requested to the flash-macrocell IP** (i.e., the design block, which includes the flash array and all of the circuitry needed to make it work). For example, in terms of memory-access time (in some cases 10 ns [in the worst-case conditions]), the best solution might not require the smallest possible 1Tr-NOR cell size: in the end, it is important to obtain the smallest possible flash macrocell area, which is not necessarily obtained with the smallest possible cell size. That is one of the reasons why, for example, a multilevel-architectural approach has never been adopted, due to the known limitations in performances such as random access time.
- **Lithography**, which in the embedded flash-process setup is driven by the reference CMOS, and the flash-cell integration should not request any particular extra feature.

As a result, while for flash stand-alone memories the involved semiconductor companies have been able to keep the 4–6 F^2 range (where F is the minimum feature peculiar of each technology node), the products with embedded 1Tr-NOR flash have seen the cell size rapidly increasing to 30 up to $\leq 50\ F^2$, with a still-reasonable trade-off between cost and performance.

The typical cross-section of a 1Tr-NOR embedded flash cell (Fig. 4.3) has remained almost the same across its evolution despite the technology scaling from 180 nm down to 40 nm and the consequent geometrical shrink in cell size.

Thus, at any given technology node, 1Tr-NOR cell sizes for embedded flash applications are mainly driven by the specifications of the products embedding

Fig. 4.3 Typical SEM cross-section of a 1Tr-NOR flash cell in the bitline direction

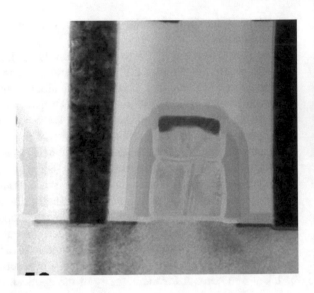

those cells. In the following sections, three different kinds of MCU products will be thoroughly analysed:

- the secure MCU
- the general-purpose/low-power MCU
- the automotive MCU

For the secure MCU, the most important flash-cell feature is the ability to ensure ≤ 1 million cycles with an erasing time in the millisecond range. That ability can be achieved by enlarging the coupling ratio between the control and floating gates, which usually leads to a larger floating-gate pitch with respect to the 1Tr-NOR flash cells used for other applications. For general-purpose/low-power applications, the minimum active pitch is targeted, because the flash cell active area width largely determines the current consumption.

For automotive products, a primary feature is an access time <20–30 ns at the end of the life, under the worst temperature conditions (with junction temperature as high as 165 °C). Access time can be achieved with a proper active width sizing, which is normally quite larger than the minimum technology feature.

Thus, despite it has not being possible to elaborate a steadfast rule linking geometrical dimensions and products, it is safe to say that general-purpose/low-power products are the most demanding in terms of cell size, whereas the 1Tr-NOR flash cell for secure applications has the largest size. In the most recent technology generations, looking at the solutions implemented at STMicroelectronics, there have been cell size differences between 20 and 30% depending on the specific technology node.

In any case, with the above-mentioned approach, and regardless of the embedded 1Tr-NOR flash application, no technology or reliability limitations are foreseen for the embedded 1Tr-NOR flash cell scaling down even to 1× nm embedded flash-technology nodes. This is because 1Tr-NOR flash cells with cell sizes <0.04 μm^2 have been put into production in recent years by stand-alone memory companies.

Nevertheless, the extra costs associated with the specific steps to shrink the cell, as well as the extra process steps needed to secure the cell compatibility with the largely increased CMOS complexity (starting at 28 nm with the high-k metal gate dielectric and then with the FinFET going to 1× node), suggest the end of the development road for the 1Tr-NOR flash cell. Therefore, different solutions for embedding NVM and advanced CMOS must be investigated: System-in-package (not exactly a newcomer in this field [10]) and BEOL non-volatile memories (such as MRAM, PCM, ReRAM) [11]) are considered more promising solutions and are gaining increasing attention.

4.2   Secure eFlash Constraints

Within the MCU market, secure MCU application requirements are distinctly different than those of general purpose or automotive applications. The foundation of the secure MCU is the ability of its hardware and software to protect information against any known type of attack. This is performed by complex encryption algorithms and identification procedures aimed at assuring a high level of tamper resistant countermeasures for exchanged data while restricting access to identified entities.

As in any other type of MCU, the e-NVM macro-cell must allow an ever-increasing amount of executable code to be loaded, which requires a sizable amount of the addressable real estate in the available memory plane, while the amount of write/erase operations are limited for this part of the memory.

In contrast, specific to the secure eFlash MCU is the need for fast identification procedures, and execution of encryption calculations requires access in reading and writing to a large amount of data type, i.e., bytes, words, or group of words, in a limited time.

That is, most of the time any access to the secure content implies data manipulation at high speed that demands for fast erase and program operations on small amount of data (granularity) and fast reading access to both data and code.

The ideal e-NVM for a secure MCU should embed the following characteristics:

1. Compact bit cell for medium code storage size in the range of a few MB (at least in the range of the present applications at 40 nm)
2. Capability to alter a limited amount of data

   (a) Byte granularity → preferred
   (b) Page granularity (64 B, 128 B, 256 B ... etc.) → considered acceptable in most cases (at least so far)

3. Fast erase and program operation
4. Low dynamic consumption

    (a) ETSI (European Telecommunications Standards Institute)-standard consumption defined for Telecom market
    (b) RF constraints for contact-less operations with minimum field

This chapter details all of the above-mentioned characteristics.

Data alterability (granularity, modify speed)

Several authentication protocols are used in secure applications, each with its own particular aspects. In most cases, it is necessary to perform a certain number of data reading/erase/write operations in a short time periods, normally in the range of few milliseconds.

Blocks of data (for example, 16 B) must be manipulated several times with elementary operations. For instance,

- Read the content
- Increment data in a byte or group of bytes
- Decrement data in a byte or group of bytes

The integrity of manipulated data must be assured even in the case of a sudden power shut-down, which may occur for any of several reasons:

- Battery off
- RF loss of field in the case of contact-less use
- External attacks

According to this scheme, the best NVM solution for a secure MCU must fit challenging requirements with important consequences for technology and design aspects.

The need to manipulate small amounts of data several times in a short time frame has spurred the on-going interest in the most popular e-NVM solution adopted so far in the secure MCU market: the byte-erasable EEPROM (2T floating-gate cell with a full Fowler–Nordheim programming mechanism). This remains valid today as well, but the increasing calls for compact cell size and silicon cost reduction have encouraged silicon providers to adopt new solutions with page-erasable capability.

In contrast, the page-erase operation does not allow the alteration of only a single byte or word, and this has many implications for application developers as well as e-NVM technology and design developers.

As mentioned previously, in the context of secure applications, information at the bytes level can be altered during subsequent steps, for example, by data increment and/or decrement operations. Each time a word or byte must be modified, in principle the entire content of the "addressed page" must be processed while assuring the integrity of data against any kind of accident. This is accomplished according to the following minimal scheme:

1. Read the content of the page where the byte is located (addressed page)
2. Write the content of the addressed page in an NVM buffer (two phases could be required)

   (a) Page erase of the buffer zone (in case it is already full)
   (b) Program NVM buffer with the content of the addressed page

3. Program the addressed page with the new content

This minimal sequence, mandatory in order to avoid loss of data, for example, in the case of tearing (power shut-down) during the modify operations, gives a first idea of the need for Write Cycling and Speed throughput, which is even more demanding in the case of page-erasable NVM architecture than for classic EEPROM architecture.

To summarize, from a general viewpoint, secure application—use cases (authentication protocols, for example) require a significant number of basic e-NVM operations to be performed on data (byte or words) in a short time frame.

However, the adoption of a page granular e-NVM imposes additional requirements linked with the need of traditional EEPROM word- or byte-alterability emulation:

- Increased cycling capability
- Increased erase and program speed (bandwidth)
- Faster access time (code execution)

Dynamic consumption constraints

Current consumption is traditionally a very sensitive feature in the secure MCU due to average and peak current constraints required from the ETSI standard for contact applications.

In addition, recent market interest in contact-less applications is driving an increasing relevance in dynamic consumption.

In both contact and contact-less cases, the key differentiating factor is the intrinsic capability of the e-NVM to perform Reading and Modify operations by consuming the lowest possible amount of energy (average current) while avoiding a peak of energy request in a limited amount of time (current peak).

For the contact case, as already mentioned, the current-consumption constraint required by the ETSI standard can be summarized according to the following rules:

- Classic SIMs → 4 mA (1.8 V)
- USIM and eSE (secure element)

  - 5 mA (1.8 V)
  - 6 mA (3 V)
  - 10 mA (5 V)

Endurance and Reliability

Various mission profiles are possible in secure MCU applications such as banking, PAYTV, and M2M, for example, with a well-identified common denominator: the request for a very high number of programming cycles, probably the highest one for e-NVM compared with other types of applications.

In most cases, as previously anticipated, the overall memory plan is used in part for code storage (few cycles), while the remaining part for data storage (high cycling demand), with a flexible allocation of the related portion of the memory, is handled according to the identified use case.

To give an idea of the possible use: In a memory size of approximately 2 MB, it is quite common to consider that only 50% of the total memory will be used for code storage, whereas the remaining part is available for data storage.

The demand can be summarized as follows (variable according to the mission profile):

- Code storage: low cycling and 10 years' data retention at $T_{amb} = 105\ °C$
- Data storage: 500-Kc page at 105 °C
- Product-failure rate <10 ppm

Of course, the increasing size of memory embedded in the secure MCU implies an impressive number of hours of HV stress, which is cumulated in the circuitry used to generate and manage all program and erase operations.

To better clarify this point, we can calculate the cumulated HV stress that must be typically sustained by the HV circuitry by a secure-element device such as the ST33J2M0 (40-nm node):

- Memory size 2 MB
- Memory allocated for data cycling (depending on the application) approximately 1 MB
- Page size 256 B (64 words by 32 bits)
- Maximum 500 K cycles/page

In this example, the number of hours the device must operate under HV conditions is calculated as follows:

$$4096\ (\text{pages}) \times 500\ \text{Kc} \times (2(\text{program}) + 2(\text{erase}))\ \text{ms approximately } 2275\ \text{h}$$

The charge pump and the overall HV circuitry must be compliant with approximately 2300 h of HV-operating conditions without fails. This necessitates a highly robust process and design, i.e., implementing advanced solutions aimed at limiting any possible failure mode while keeping a high level of performance (speed, robustness, and small area).

4.2.1 Page-Erasable Secure NOR eFlash

Flash cell-specific characteristics

Secure MCU applications demand fast erase and program times due to the high number of write operations performed in a reduced time slot. As mentioned in the previous example, an erase operation on the order of 2 ms typical is normally required, while at the same time the Program operation must be fast enough to fill the content of the page again after erase in a similar amount of time (approximately 2 ms). In addition, the current-consumption budget is also limited in both of the operations.

In contact applications (ETSI standard for 1.8 V range), the overall current budget for the entire device during Modify operations is 4 mA, whereas the current allocated for the e-NVM is <3 mA. In the end, the designer faces two opposing demands:

* Fast erase/program throughput
* Low dynamic consumption

The 1Tr floating-gate cell normally used in a sector-erasable flash is not conceived to fulfill these opposing requirements:

* The erase operation is in the range of some tens of ms.
* The program operation, even if fast, intrinsically consumes a high amount of energy due to the channel hot-electron (CHE) programming mechanism.

Due to those reasons, the 1Tr-NOR flash cell used in page-erasable architecture must be significantly different from the standard in at least two characteristics:

* **Fast Fowler–Nordheim**
* **Efficient programming** (CHE with the lowest possible peak current)

Fast Fowler–Nordheim

As in the sector-erasable solution discussed previously, the page-erasable flash cell is erased by the Fowler–Nordheim tunneling effect. Bias to the gate and bulk terminals is applied according to the classic scheme shown in Fig. 4.4 where reported voltage values are purely indicative.

According to this scheme, there are two ways to obtain fast erase operations:

* Increase the erase voltages
* Increase the control-to-floating gates coupling factor

Increasing the erase voltages is not a viable solution due to the technological side effects in terms of flash cell and HV CMOS reliability. For instance, cell aging is accelerated if too-high voltages are applied during cycling by inducing fast degradation on tunnel oxide and charge trapping with detrimental effects in terms of endurance [12]. In contrast, HV CMOS transistors should be implemented with thicker gate oxides and longer minimum lengths, which would imply

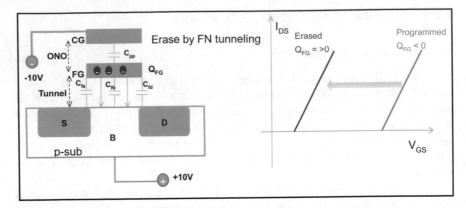

Fig. 4.4 Erasing-bias scheme for 1Tr-NOR page-erasable flash

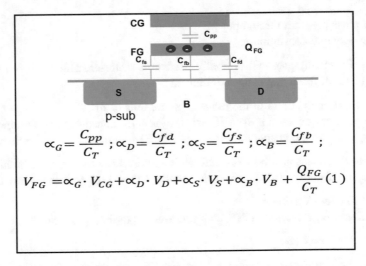

Fig. 4.5 1Tr-NOR flash cell-coupling factors

low-performance HV devices with strong limitations in terms of speed and increased circuit periphery areas. Row decoders and column decoders especially would be affected [13]. In other words, this solution cannot be put into practice without affecting overall performance, reliability, and cost.

However, considering the principle on which the FN-tunneling mechanism is based, it is quite obvious to enhance erase efficiency by increasing the coupling factor (see Fig. 4.5 as a reminder of the coupling factor role).

Increasing the coupling factor means increasing α_G, which can be achieved due to enlarged flash-cell "wings," i.e., the overlap area between the control and floating gates over the field oxide. The resulting drawback is a larger-sized cell with respect to a standard 1Tr-NOR flash at approximately 20% additional area, but the advantages in terms of performances and reliability are enormous.

Efficient Channel Hot Electron (CHE)

Page flash is programmed by the same physical mechanism of standard 1Tr-NOR (CHE).

The major difference is in the need for optimal power consumption *versus* speed, which can be obtained by proper design and technology choices. To highlight the basic bias scheme used, the well-known mechanism for this is reported in Fig. 4.6.

Programming efficiency is judged by the capability of changing the status of the cell from low threshold to high threshold while consuming the lowest possible energy. For a 1Tr-NOR flash, this target can be obtained with three concurrent strategies:

- Increase of flash-cell channel-implant dose
- Decrease of drain voltage bias
- Use of programming box pulses strategy

In more detail, this means that a higher cell-channel implant dose makes possible an improved CHE programming efficiency while reducing the current peak on the drain of the cell. The drawback of this choice is a potentially increased drain leakage, which can be contained with an optimized drain voltage bias.

Lowering the drain voltage as much as possible has two major advantages:

- Reduced leakage consumption in program operations due to the unselected cells belonging to the selected bit line
- Optimized drain current consumption on the selected cell

In Fig. 4.6, the values +HV and +MV range, respectively, between [8–10 V] and [3.5–4.5 V] depending on the technology node and the intrinsic cell characteristics. The third key element in the reduction of current-consumption strategy during programming is proper selection of the programming-pulse approach.

As will be shown in the following text, the box-pulse method can be used to obtain a drastic reduction of the current consumption needed by all of the circuitry actuators including, but not limited, to Charge Pumps and DAC down to a level that allows a 1Tr-NOR flash cell to be used within the most demanding dynamic consumption perimeters—the ones dealing with RF applications.

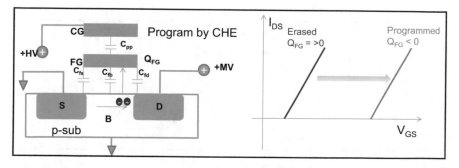

Fig. 4.6 Programming-bias scheme for 1Tr-NOR Page flash

Disturb issue

To introduce the peculiar and intrinsic disturb mechanism present in page-erasable 1Tr-NOR flash architecture, the basic sector organization and the bias scheme needed to perform modifying operations (erase and program) will be examined.

In Fig. 4.7, the erase operation is described addressing one page (or row), and the bias scheme used to allow the Fowler–Nordheim (FN) mechanism will be effective only on the selected page.

The selected page (named WL <0> in the figure) is biased to a proper negative voltage (values are purely indicative), whereas the unselected pages are all biased to 0 V. The bulk terminal is biased to a positive voltage as was performed in the sector-erasable NOR flash discussed earlier in this chapter.

Although the bias is such that the tunneling effect is activated in all of the cells belonging to the selected page, on the unselected pages this does not happen because of the reduced net voltage. However, it is clear that reiterated erase operations on the same or other pages may induce a kind of soft-erasing effect. This disturb, cumulative over time, could induce a charge loss in the programmed cells, the threshold of which may shift lower than the read voltage (RV), thus causing loss of the information. This type of disturb is called "bulk-erase disturb" and requires appropriate design countermeasures (see Fig. 4.8 for reference).

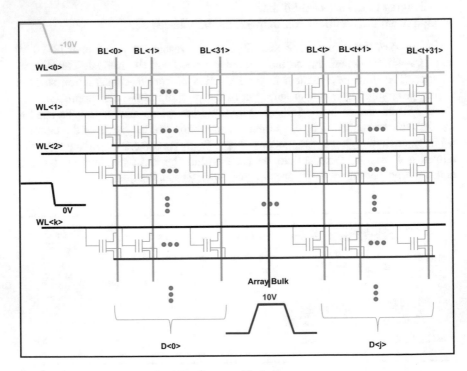

Fig. 4.7 Bias scheme for a 1Tr-NOR page-erasable flash

Fig. 4.8 Bulk-erase disturb
mechanism

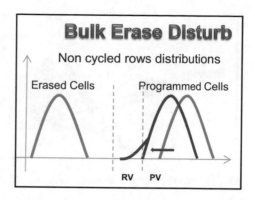

Tearing issue

In all secure MCU applications, and especially in RF, a sudden power shut-down
(tearing) must not prevent the device from working properly at the next power-on
reset. The integrity of sensitive information contained in the memory must be
maintained whatever the interrupted operation. This is a key requirement that in
theory may prevent the use of 1Tr-NOR page-erasable flash in secure and RF
applications.

This limitation descends directly on one hand from the absence of a select
transistor and on the other hand from the way the erase operation is performed in
any 1Tr-NOR flash. As previously mentioned in this chapter, the erase operation is
a combination of two different phases:

1. FN-erase phase
2. soft-programming phase

The first consideration is due to the flash sector architecture, wherein the minimum group of cells that can be erased at the same time belongs either to a page or
row, which normally contains thousands of cells.

The fn-erase phase pushes the threshold voltages of all of the cells belonging to
the page well lower than the so-called erase-verify (EV) voltage, which represents
the highest threshold voltage value for the erased cells to be identified as "erased"
(logic 1) by the reading operation.

However, it often happens that part of the distribution of the erased cells goes
lower than the so-called depletion-verify level (DV). The DV level is the minimum
threshold voltage for all of the erased cells belonging to the sector, below which a
cumulative leakage could be present along the bit line and prevent a correct reading
operation.

The soft-programming phase is a dedicated and delicate soft-programming algorithm, which corrects the erased-cell threshold by applying a controlled-programming
operation in order to position the cells above the just mentioned DV level. If an erase
operation is interrupted before the completion of the soft-programming phase, cells

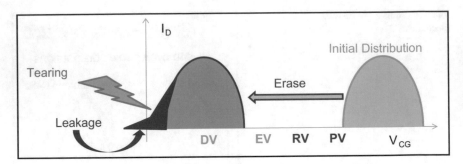

Fig. 4.9 Possible memory content-corruption scheme during erasing operation

with threshold lower than DV may generate a potential leakage in the flash sector, which can cause corruption of the sector. This mechanism is shown in Fig. 4.9.

Summary

To recap what has been described thus far, the 1Tr-NOR page-erasable flash has some intrinsic limitations that could prevent, in principle, its use in a secure MCU. Those limitations are mainly these three:

1. Current consumption: Program operation requiring high current consumption under medium-voltage conditions (some tens of μA at MV applied on the cell drain)
2. Bulk-erase disturb: Reiterated erase operations may induce a shift of programmed distribution with loss of information and content corruption
3. Tearing issue: Power shut-down during the erase operation may induce leakage if the erase operation has not been completed with the soft-programming phase

In the following paragraphs, an example of the typical implementation of a proper memory architecture and dedicated design solution, as set up at STMicroelectronics, illustrates how all of the above-mentioned limitations have been surmounted, thus enabling 1Tr-NOR Page-Erasable flash to be used in the most advanced secure MCU devices present today on the market.

4.2.2 Secure e-Flash Design Architecture

Top-Level Block Diagram

A Secure 1Tr-NOR Page-Erasable flash macrocell block diagram as implemented by STMicroelectronics is illustrated in Fig. 4.10. Hereinafter, a short description of the most important blocks is provided.

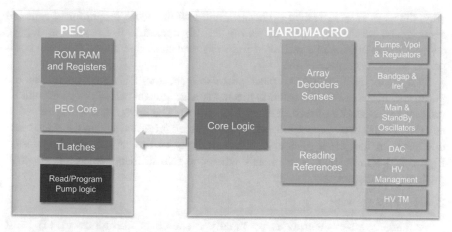

Fig. 4.10 Block diagram of 1Tr-NOR page-erasable flash macrocell as implemented by STMicroelectronics

The 1Tr-NOR eFlash macrocell is organized into two major sub-systems:

(a) program-erase controller (PEC): A full custom digital controller executing the embedded algorithms used to generate program, erase, and test sequences
(b) Hard macro: The analog part of the memory including the array and all of the sensing and programming actuators: charge pumps, references, regulators, DAC and sense amplifiers, etc.

Tearing and Disturb Management

Both tearing and disturb management are key to overcoming the intrinsic limitations of the 1Tr-NOR cell. There are two concepts behind the implementation:

1. The refresh of programmed cells to overcome the bulk-erase disturb
2. Tracking the status of the erase operation to overcome the tearing issue

Both concepts can converge in a competitive architecture.

Disturb Management

The way to overcome the cumulative effect of threshold shift on a programmed cell is to perform dedicated refresh operations. A refresh operation is nothing more than a re-programming operation performed on a cell that is moving lower than the programming-verify threshold.

In other words, the erase-operation sequence can be organized in a way that includes not only erase and soft programming but also a phase of reading operations that checks the threshold of programmed cells and performs a re-programming of those that are losing electrons from the floating gate. Of course, this operation must be implemented with a rotation mechanism.

In practice, at each new erase operation, the address of the page that must be refreshed is recovered from an NVM buffer, which stores this information. In this way, the page to be refreshed is checked, and low-threshold programmed cells are re-programmed. At each new erase, the refresh operation is executed at a page address following the one refreshed in the former erase operation. This step comprises the rotation mechanism.

In a sector made up of 512 pages, for example, after 512 erase operations are performed in the sector, the rotation mechanism assures that all of the pages will be refreshed (periodicity of refresh every 512 erases). The refresh can be organized in such a way that two or more following pages can be refreshed within the same erase operation. In that case, of course, the periodicity of the refresh will be shorter. Figure 4.11 shows the effects of bulk erase and refresh management on a typical programmed distribution.

Tearing Management

As already stated, tearing is potentially a source of corruption in 1Tr-NOR because the soft-programming operation could have been interrupted at the power shut-down. To prevent this issue, the following information is needed:

(a) At power-on, it must be recognized that an erase operation was interrupted indicating that a leakage is potentially present in one sector
(b) The address of the sector, as well as the page in the sector, affected by the potential leakage must be rapidly available, meaning that in practice a pointer to the leaky row must be provided
(c) A cleaning operation of the leaky row must be executed in order to remove the leakage, which involves an execution of a repair operation on the row identified as a potential source of leakage

Fig. 4.11 A typical example of flash-programmed cell distributions after bulk erase and following refresh application

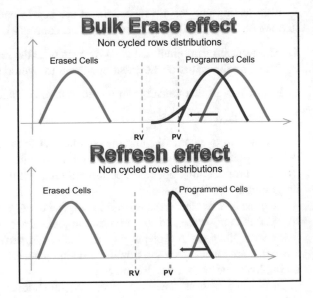

In order to obtain the above-mentioned three essentials, STMicroelectronics adopted a solution of introducing additional special rows in each sector of the e-NVM [14]. These rows are called "manager rows" (MR), and their function is to track, step by step, all of the various phases of the erase operation with a special coding of the MR bits. In practice, each time a row in the sector must be erased, the bits of the manager row in use are updated in order to provide the following information:

1. The address of the page in the sector that is under erase
2. The address of the page on which the refresh operation must be executed
3. The status of the erase operation

    (a) Manager row status (filled, empty, or under erase)
    (b) Refresh (executed or not)
    (c) Erase (executed or not)

This is the complete set of information needed to perform both tearing and refresh management.

In case of a tearing occurrence causing the interruption of an erase operation, at the following power-on sequence the PEC of the e-NVM will execute a fast check of the status of the manager rows inside each sector, recognizing (in a few μs) if one of the sectors has been affected by power loss in the erase. This phase is called "get repair".

In case such an event is detected, all of the above-mentioned information is provided at top level to the product CPU. Then the repair sequence is run directly targeting the row affected by the potential issue. All of the bits of the row are checked in order to identify leaky bits—i.e., those bits lower than DV level—and remove the leakage through soft programming of the leaky bits. Figure 4.12 shows the sequence of the get-repair function.

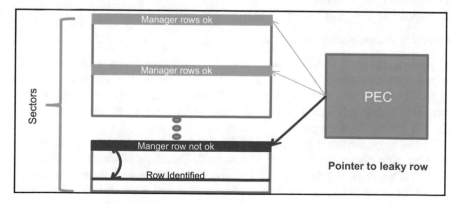

Fig. 4.12 The get-repair sequence

Current-Consumption Optimization

This section focuses on the methodology used to minimize current consumption in program operations. Current consumption is the major drawback related to 1Tr-NOR flash because it is well known that the CHE-programming mechanism is potentially quite demanding in terms of energy consumption.

The voltages used to activate CHE were highlighted previously, and Fig. 4.6 provides details on the channel hot electron-programming mechanism. The current provided on the drain of the cells is in the range of tens of µA, and the drain voltage must be kept at approximately 4 V. Due to the low-supply voltage-operating conditions (approximately, minimum 1.0 V of internal voltage), it is obvious that such an amount of current must be generated by a charge pump.

A charge pump has its own inefficiency, $\eta = I_{VCC}/I_{OUT}$, which is dependent on the following parameters:

- Minimum supply voltage (V_{CCMIN})
- Output-regulation voltage (V_{OUT})
- Number of stages (N)
- Pumping stage-capacitance inefficiency ($\alpha = C_{PAR}/C_{STAGE}$)
- Here C_{PAR} represents the parasitic capacitance in parallel between the bottom plate

$$\eta = (N+1) + \alpha * \frac{(N^2 * V_{CCMIN})}{((N+1 * V_{CCMIN} - V_{OUT}))}$$

and the bulk of the stage capacitor C_{STAGE}

Clearly, any charge pump amplifies the consumption required by the programming operations due to its intrinsic inefficiency.

A twofold strategy is used to optimize programming consumption:

1. Adoption of box pulse
2. Improving design efficiency

Lowering Programming Consumption

Historically, the 1Tr-NOR flash CHE has been implemented with two possible approaches: the ramped-gate pulse and the box pulse. the ramped gate-pulse approach was published in a 1999 paper [15] and was aimed at controlling the drain current of a flash cell during program operations by applying a ramp on the gate of the cell during the drain pulse.

In this way, the cell's overdrive is kept constant during programming, and the programming current remains constant as well. The slope of the ramp allows the current of the drain and the programming speed to be controlled: In other words, the higher the ramp slope, the higher the consumption and the speed and vice versa.

Although this approach, extensively used in many applications, is in principle well suited to control the consumption of the cells, in practice, for the secure MCU it brings with it two major drawbacks. The first is that the duration of the programming pulse must be in the range of several microseconds if a low current consumption is to be obtained, thus inducing a reduced programming bandwidth. The second is the high consumption induced by the ramp: In fact, the ramp on the gate is applied by a row decoder, which represents a load capacitance in the range of tens of pF for the ramp generator. This is built with a DAC supplied by an HV charge pump. Due to the fact that the DAC must generate a ramp in the range from 3 to 9 V in a few µs, this induces a current consumption from the HV pump in the range of tens of µA.

Because of the multiple stages needed to generate voltages as high as 10 V, the HV charge pump is so inefficient that the consumption due to the ramp generation is comparable with that needed to supply cell drain during program pulse, thus almost doubling the global current consumption in this phase because drain pulse and ramp are applied at the same time.

Ultimately, the sum of the two currents can reach up to 2 mA, quite a high current peak value for a secure MCU application where the current peak amplitude and duration is a key parameter (both for ETSI and RF use cases).

The box-pulse approach, in contrast, entails the application of a constant high voltage on the gate while the voltage on the drain is applied. This has two major advantages. The first is that there is no longer a simultaneous superimposition of the two current contributors. In fact, the current consumed to bias the gate of the cells (coming from the DAC and the HV charge pump) will no longer be applied at the same time as that to the drain current, such that the two contributions will be not summed at the same time (reduction of current peak). The second advantage is the duration of the programming pulse. Each programming pulse in the box approach will have a duration in the range of 1 µs, thus increasing bandwidth in the program by a significant factor.

Overall, the box-pulse approach is the optimal solution for secure MCU applications with 1Tr-NOR flash because it allows improved bandwidth and reduced consumption *versus* the ramp approach (see Fig. 4.13 for a comparison).

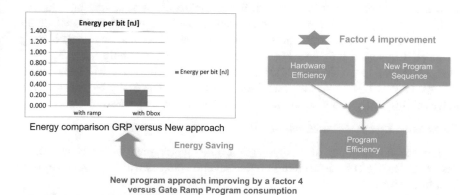

Fig. 4.13 Programming box-pulse improvement wrt-ramped pulse as implemented by STMicroelectronics secure MCU

The second element to obtain competitive consumption is the design-efficiency optimization. The focus here is on two main examples in the design: charge pump and DAC.

Charge-Pump Design

First, here is a short introduction of some unique characteristics of the secure MCU before discussing the *Charge-Pump* design.

In secure applications, the external V_{CC} can have a wide range from 1.65 to 5.5 V. In addition, there is the need for a robust design against external attack—for example, a rapid change of the external supply—and avoiding a direct connection of the charge pump to the external V_{CC} is also preferred.

In many cases, this means that the NVM charge pump could be directly connected to the internal V_{DD} of the product, which is normally generated through a dedicated DC/DC converter to be compatible with a high-performance embedded CMOS. However, when this internally generated supply voltage is too low—and in recent technology, a node can be as low ≤ 1 V—the best way to proceed is to generate a second internal voltage in the range of 1.45 V, which is suitable for obtaining good efficiency. This requires a dedicated DC/DC converter for the charge pump.

For a more precise idea, assume the hypothesis that a charge pump able to generate a V_{OUT} of approximately 4.5 V with a driving capability of 800 μA is needed. If the charge pump is supplied by 1 V and $\alpha \sim 0.15$, then the optimum number of stages to obtain the minimum consumption is $N = 5$, and the associated inefficiency is $\eta = 8.5$. In contrast, if the charge pump is supplied with 1.45 V with the same α parameter, the optimum number of stages to obtain the minimum consumption is $N = 3$ with an associated inefficiency of $\eta = 5.51$. The difference in terms of efficiency is >50% less in the second case, and this provides a great advantage, since the current consumption is reduced by the same amount. In addition, the higher voltage used to supply the Charge Pump gives an advantage, in terms of area, of a factor 2 versus the 1 V supply. Finally, the strategy used to reduce the area of the pump is to push the pump frequency to a very high range.

In this example, STMicroelectronics implemented the use of a special pump solution with LV MOS transistors as switching elements and with self-biasing CASCODED stages [16] (see Fig. 4.14). Consequently, a supply voltage of the charge pump up to 2 V (if needed) is possible even though the LV MOS maximum voltage range is 1.32 V. At the same time, the frequency can easily be pushed in the range of 200 MHz with a strong reduction of both area and consumption.

Continuous Mode DAC with pure capacitive-feedback network

The second example of design optimization worth mentioning is the DAC design. In flash, a DAC is used to generate several voltages with high accuracy and high bandwidth.

Because high voltages are needed, the output stage of this circuit is supplied by an HV charge pump with high intrinsic inefficiency, and $\eta \sim 20$ is quite common.

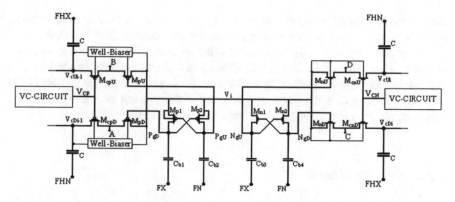

Fig. 4.14 Scheme of an optimised charge pump implemented by STMicroelectronics

This means that the DAC must be an amplifier working with extremely low biasing current, such as approximately 2 μA, and at the same time be able to generate an output current that can be as high as 200 μA—not a trivial specification at all, in fact, with two orders of magnitude between output and static currents.

Concurrently, the static-current consumption is determined by the feedback network, which is normally implemented by a resistive divider. In order to reduce the static-current consumption, even when 10 V must be generated, the resistive network would be in the range of 10 MΩ. This value represents a problem because

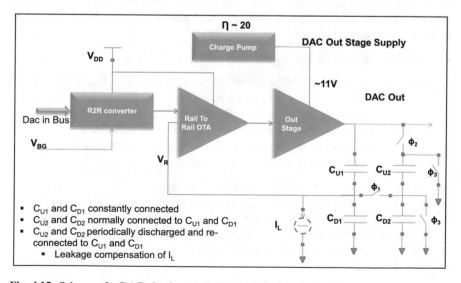

Fig. 4.15 Scheme of a DAC circuit as optimized by STMicroelectronics

the associated resistance would occupy a large silicon area and would have intrinsic parasitic coupling capacitance that might reduce the bandwidth of the DAC.

The adopted solution [17] is a DAC that implements a fully capacitive feedback network. This type of feedback has several advantages—a fast bandwidth and theoretically zero static consumption. The need to compensate for the leakage of the feedback net induces a static consumption that is not zero but is as low as 200 nA, which represents outstanding performance (see Fig. 4.15).

4.2.3 Silicon Results

The 1Tr-NOR Page-Erasable flash is already present in many of STMicroelectronics' secure MCUs. Two types of applications, RF and contact, are discussed.

RF Product Application

A high-performance version of this architecture has been implemented in STMicroelectronics' **SR31G480**. This product embeds a 480-KB page-erasable flash with outstanding erase and program speed performances because of optimized page-length organization (64 B/page with highly efficient row-decoder architecture):

- Typical page-erase time of approximately 1.4 ms
- Typical page-program time of approximately 0.7 ms
- Cycling capability of >500 Kc/page
- Minimum RF field of H_{min} <0.9 A/m

A current consumption <1.1 mA of the e-NVM macrocell in an RF configuration provides the best in-class H_{min} value. This product is available in an 80-nm technology node.

Contact Product Application

At the end of 2015, STMicroelectronics announced the introduction of the ST33J2M0, a 32-bit secure MCU for telecom applications. The device is diffused in 40-nm embedded flash technology node and implements a 2-MB embedded 1Tr-NOR Page-Erasable flash. The page length (256 B) is optimized to obtain the best compromise between silicon area and electrical performances. Some of the characteristics of this product include the following:

- Typical page-erase time of approximately 1.8 ms
- Typical page-program time of approximately 1.8 ms
- Cycling capability of >500 Kc/page
- Frequency of approximately 70 MHz
- ETSI consumption-requirement compliancy

4.2.4 Conclusions

The 1Tr-NOR Page-Erasable flash is an embedded solution with outstanding performance suited for secure MCU applications. The intrinsic limitations of the 1Tr-NOR flash cell are able to be overcome with proper architecture and dedicated design solutions. The result is a high-performance NVM macrocell with fast erase- and program-operations capability and low dynamic consumption combined with the well proven high-cycling robustness of the 1Tr-NOR flash.

4.3 Low-Power Embedded 1Tr-NOR Flash Macrocells

4.3.1 Introduction

Until recently, the most important specification parameter for an embedded NVM macrocell within a MCU was random-access time. Consequently, the memory architecture definition and the implemented design solutions were mainly oriented to satisfy this requirement.

Now, as previously discussed in the chapter on secure MCU, the power-consumption features, and particularly the read dynamic current consumption as well as the static consumption, are replacing random-access time as key parameters in the demand of the microcontrollers linked to the increasing market of portable electronic device applications.

This increasing demand has prompted MCU manufacturers to develop and implement design and technological solutions not only oriented to the reduction of device current consumption during operations but also when the device is put in a non-operating mode. These requirements must be taken into account in the design of the 1Tr-NOR e-flash macrocells for low-power application MCUs.

In order to define the main consumption specifications, the most common device-operating scenarios must be defined. There are three generic key operating modes:

- Stop (deep power down)
- Idle
- Read

When analyzing the design-conception challenges, it must be reiterated that the 1Tr-NOR embedded flash macrocell is characterized by having an array decoder that must manage high voltages (HV) during he erase and program operations. Therefore, it follows that either the wordline or the column decoders are built using HV MOS devices.

The problems related to the voltage stress applied to those devices is not limited to reliability issues (e.g., oxide breakdown) but also to degradation of the device's characteristics due to the cumulative electrical stress applied to those devices during the HV operations. It is well known that the application of an electric field above a certain electric field induces MOS characteristics variations such as threshold voltage drift and transconductance degradation.

Those degradation mechanisms, which become enhanced at high temperature, must be carefully evaluated during the macrocell design phase in order to avoid device-performances degradation all along the product life span.

A possible worst-case scenario can be identified in the wordline decoder, which must be immune to any significant selection-speed degradation that could impact read-access time on highly cycled devices. In general, the requirement of HV MOS increases the level of difficulty in the design of compact and fast wordline and column decoders of a 1Tr-NOR embedded flash. That is particularly relevant when there is a need to limit the dynamic consumption associated with wordline and column selection and de-selection during read operations while also taking into account that charge pumped voltages are used to supply the array decoders.

Despite those built-in disadvantages, 1Tr-NOR e-flash macrocells can achieve excellent performances both in terms of random-access time as well as with respect to the overall dynamic current consumption in read mode. A more extensive discussion of the above-mentioned three key operating modes follows.

4.3.2 Operating Modes of 1Tr-NOR Embedded Flash

Stop

When the flash is in stop mode, the current consumption must be minimized, ideally limiting it to only device leakage (junction-reverse current, MOS sub-threshold current, GIDL, gate leakage, etc.). Most of the circuits, either analogue or digital, must be switched off. To guarantee the internal power-supply level even during the stop phase, a reference voltage (band-gap reference) and the internal DC–DC converter for power-supply generation must be kept active. Special care must be dedicated to the design of those elements that play a major role in the definition of the stop static power-consumption performances.

Minimizing circuitry leakage is achieved by decreasing the power-supply level during the stop phase down to the level that guarantees the flip-flops data-retention conditions. Moreover, in stop mode part of the circuitry can be disconnected from the power supply to remove the associated leakage contribution. If the power supply and its voltage reference are provided by the microcontroller system, the entire embedded flash circuitry can be switched off. In stop mode, sub-µA stand-by current consumption can be achieved at room temperature.

Idle

When an exit from stop mode is requested, a 1Tr-NOR flash macrocell requires some time to turn on the analogue circuits and establish suitable conditions for the correct operations execution. During this "wake-up" phase, all of the analogue references (current generators, read-references voltages and currents, charge pumps) must be turned on, and the related output values must reach their steady state. This means that all of the internal references required to perform the read operations (voltage, current, time) must be ready and stable, and such conditions must be sustained with minimum power consumption.

In the idle mode, the MEMORY must be ready to execute read operations (e.g., code fetching) with low, preferably zero, time latency. Wake-up (from stop mode to idle mode)—latency time as low as a few μsec must be achieved.

In a 1Tr-NOR flash, the accuracy of the read operation is linked to the precision of the voltage level applied to the control gate (row) of the cells of the selected wordline. This voltage is generated by a dedicated charge pump, and it is mandatory to reach the regulation level before starting any read operation. Therefore, the read-charge pump supplying the row decoder during the read operation must be able to provide relevant output current when executing continuous Read operations at high frequency, whereas it must be able to sustain the required voltage level with a minimum consumption when the device is in idle mode. Usually, two different clock frequencies are used, and the charge pump must be able to quickly switch from a low consumption, low-frequency mode to high driving capability and high-frequency mode as soon as a read operation is requested.

A significant part of the idle consumption is due to the RV-regulation system. This regulator can be relatively slow when the flash is in idle mode (low clock frequency of approximately 1 MHz), but it must be much faster to react when the charge pump is controlled at higher frequency (>100 MHz) during read operation.

The sensing operation also requires a reference current, which is used to discriminate between zeros and ones. In 1Tr-NOR flash memories, such a current is often obtained using memory cells as reference elements. The "reference cells" can be configured at a suitable threshold voltage during the device test flow. The current sunk by those cells is on the order of several μA, which prevents a limit on the static-current consumption associated with the reference cells biasing. In contrast, it is difficult to effectively turn on–off the sensing current-reference circuitry at every read operation due to the high capacitive load of the transistors mirroring the current on each sense amplifier (Fig. 4.16).

A solution to reduce static-power consumption of analog reference circuits is based on the following procedure:

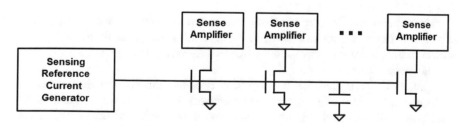

Fig. 4.16 Reference current-sensing scheme

- store the analog voltage level in a capacitor,
- switch on the generation circuitry during the short period needed to reach the steady state, and then
- connect the output to the capacitor to "refresh" the voltage level before disconnecting and turning off the circuit.

This decreases the static-current consumption according to the off–on time ratio (duty-cycle).

This procedure can be applied to other analog circuits having a purely capacitive load and that are requested to be constantly active in idle mode [18]. Thanks to these specific design solutions, idle-current consumption <10 μA can be achieved in a 1Tr-NOR e-flash macrocell.

Read

Read-operation performances are mainly defined by the access time and the dynamic read-current consumptions. The analysis of the read dynamic-current budget identifies the following main contributors:

- Row decoder
- Column decoder
- Sense amplifier
- Logic (address, pre-decoding, outputs)

In a low-power 1Tr-flash macrocell, the minimization of the dynamic read-current consumption represents a major challenge for the designer. In fact, as already mentioned in the chapter on secure MCU, generating the voltage levels required to drive the wordline decoder as well as the column-decoder pass transistors has a relevant affect on the overall read dynamic-current consumption due to inefficiency of the charge pumps.

The cost of the pumped voltages for the design of the macrocell becomes particularly relevant when the device must be tolerant to a wide voltage supply range, which makes mandatory the use of an increased number of pump stages.

The 1Tr-NOR flash memory requires a continuous selection and de-selection of the row at each read operation, even when operating on the same wordline (sequential access read) to minimize the gate stress time on the selected row's cells. And while the supply voltage of the wordline decoder is imposed by the 1Tr-NOR flash cell characteristics, the wordline capacitance can be adapted through a suitable choice of the memory-array organization.

The column-decoder pass transistors can be left selected after the end of the read operation, but this does not bring any relevant advantage in terms of dynamic consumption, since it is very unlikely that two following read operations will occur on the same column.

Unlike the procedure for the wordline decoder, the column decoder-supply voltage can be chosen by compromising between dynamic power consumption— i.e., lower voltage—and column-decoder pass transistor impedance, which affects the bit line-charging time and therefore the read-access time.

Because of the associated gate capacitance, the width of the column pass transistors plays a major role in the definition of the read dynamic-current consumption. A smaller width induces a lower consumption at the cost of an increased bit line-charging time in read operation. Moreover, for both the wordline and the column decoder, part of the consumption can be related to the transition of pre-decoding signals, which are also supplied by the pumped voltage levels. The minimization of the associated consumption can require a suitable transistors sizing, thus limiting the opportunities to speed-up the wordline and column decoding.

A relevant part of the dynamic-read current, including the capacitive current required to charge the bit lines at the suitable voltage level, is handled by the sensing circuitry. In read operation, the bit line capacitance-charging current can represent the most relevant part of the global current in the sense amplifiers.

The bit-line voltage in read operation is usually kept at <1 V in order to avoid dangerous cumulative drain stress, which could determine soft-programming errors due to CHE injection. In a low-power 1Tr-NOR flash memory, it is possible to limit the bit line-biasing voltage level during read operation, usually in the range of 400–600 mV, to reduce the associated charging current. It comes with a compromise, though: a sensing-time degradation due to the cell-current reduction. Of course, both the dynamic consumption and the sensing time are dependent on the bit-line length (capacitance).

It is unavoidable that the efforts required to reduce read dynamic consumption limits the design choices aimed at reducing the 1Tr-NOR flash memory random-access time. Usually, the access time granted by the low-power flash could represent a limit to the speed-up of the MCU system clock.

A commonly used approach for overcoming the limits of the flash read speed is to improve read-data throughput by enlarging the read parallelism. That can be extended far above the size of the word used by the MCU, and this approach can give relevant advantages in case of sequential read operations with a suitable number of clock-pulses latency.

For example, a 32-bit MCU can embed a 1Tr-NOR flash memory having 128-bit read parallelism where the duration of each read operation lasts less than four clock periods ($t_{acc} < 4\ T_{ck}$). Such an architecture is normally supported by a buffer RAM (cache RAM).

Although an instruction fetch usually requires four clock periods, in the case of sequential read (linear code) the architecture takes advantage of the anticipated read operation of the next instructions. This avoids introducing latency clock periods when the MCU fetches instructions predictively downloaded from flash into the RAM. This is known as "pre-fetch."

Moreover, pre-fetch architecture makes it possible to keep in the RAM the part of the code previously downloaded from the flash. That introduces further opportunities to speed up the code access when the MCU requires instructions already in the RAM. The block scheme of 1Tr-NOR flash memory with cache RAM is shown in Fig. 4.17.

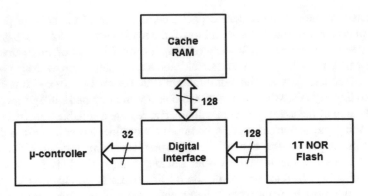

Fig. 4.17 1Tr-NOR flash memory with cache RAM schematic block

The approach discussed above means that the sensing circuitry must be able to react when an instruction that is not stored in the RAM is required, even while the sensing operation is running at a different address location. In this case, the sense amplifier must be able to suspend the current read operation and immediately restart a new one at the required location, thus avoiding the introduction of further clock cycles-access latency. Moreover, the management of the "interrupted" read operation must be taken into account in the evaluation of the driving capability of the charge pumps supplying the row and the column decoders.

The increased read parallelism can bring further advantages when the flash embeds error-correction codes by allowing a reduction of the number of required control bits, with the following optimization in terms of memory area and average dynamic consumption, while still maintaining satisfactory correction coverage.

From the viewpoint of read-dynamic current consumption, the increased read parallelism has an advantageous effect, because, despite the part of the current budget associated with the sensing circuitry grows linearly with the number of sense amplifiers, the current required by the row decoder remains independent of the read parallelism, and the consumption due to the column decoder is only partially increased by the higher number of selected pass transistors.

4.3.3 The Sensing Operation

The design of the read path, and particularly the conception of the sensing circuitry, represents one of the major challenges of a low-power 1Tr-NOR flash architecture definition.

It has already been mentioned that most of the current consumption by the sense amplifiers is due to the bit-line pre-charge current. It follows that, assuming a read parallelism on n bits and a bitline capacitance C_{BL}, the minimum dynamic sensing-current consumption achievable cannot be lower than

$$n \times C_{\mathrm{BL}} \times V_{\mathrm{BL}}$$

per read-frequency unit where V_{BL} is the bitline pre-charge voltage.

Reduction of the bitline capacitance would improve both dynamic consumption and access time, but the required array partitioning implies a trade off with the impact on silicon area.

With an appropriate cell layout, the 1Tr-NOR flash shows the advantage of proper cell transconductance. Therefore, in read operation, the bitline voltage can be reduced as long as the cell current remains sufficient to guarantee reliable sensing operations compliant with the required access time.

In order to optimize the read-dynamic consumption, it is an option to aggressively reduce the bitline voltage during read operation dealing with low cell currents and consequently with low-sensing reference current. Choosing this option necessitates adopting sensing schemes that are particularly robust in terms of accuracy, precision, and noise immunity.

Although a single-ended sensing architecture can be suitable for fast access time 1Tr-NOR flash memories, a balanced double-ended sense amplifier is often a privileged choice for ultra low-power applications. If double-ended sensing is chosen for a 1Tr-NOR flash, the memory architecture is usually structured as shown in Fig. 4.18.

This structure provides a balanced system where the capacitance of the selected bitline (read side) is compensated through the selection of the symmetrical one, which plays a role of capacitive load because it has no row selected (reference side).

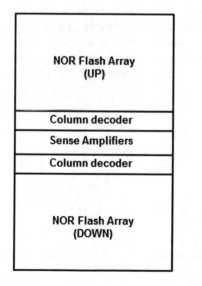
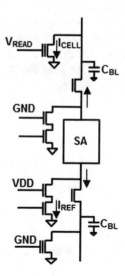

Fig. 4.18 Flash-memory architecture with double-ended sensing scheme

Concurrently, the sensing reference current is applied on the sense-amplifier reference input.

One of the most critical elements of the 1Tr-NOR flash sense amplifier is the bitline pre-charge circuit, which has the "duty" of bringing the selected bitline to the required voltage level for the cell-current evaluation. A simple mechanism for bitline pre-charge is based on the use of an n-channel MOS-source follower having the gate biased at a suitable voltage level as depicted in Fig. 4.19a. An advantage of this widely used approach is that on the drain side of the follower, the cell current is observed in a low-capacitive node.

The drawback of this simple scheme is the length of time required to obtain a satisfactory bitline pre-charge and the consequent fading of the capacitive current flowing into the follower. A solution adopted at STMicroelectronics to improve the bitline pre-charge time is shown in Fig. 4.19b [19].

The gate of the source-follower stage is de-coupled from the reference voltage through impedance (a resistor). That makes the gate of the follower susceptible to coupling with the source and drain nodes, thus generating a self-boosted mechanism at the circuit turn-on due to the MOS drain/gate and source/gate capacitances. The circuit can be optimized through a suitable sizing of the follower MOS and the de-coupling impedance. This scheme's advantage is that the reference-voltage generator drives a purely capacitive node although it requires a large capacitance to minimize the perturbation of its reference voltage level.

Another widely used scheme for bitline pre-charge is shown in Fig. 4.20a. In this scheme, the output of an inverter stage drives the gate of the n-channel follower. This bitline pre-charge approach takes advantage of the feedback to guarantee a fast pre-charge mechanism, but the bitline pre-charge level remains linked to the trip-point of the inverter stage. The limitations of this scheme are a dependence on the bitline level by the supply voltage and a susceptibility to supply noise. Moreover, the inverter could be not compatible with a low-power supply level. Nevertheless, because the 1Tr-NOR flash memory can rely on the external supply —which is higher than the internal one—the inverter stage can be controlled as shown in Fig. 4.20b.

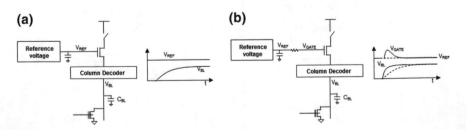

Fig. 4.19 Example of flash macrocell bit line-precharge scheme (**a**) and one-ST implementation (**b**)

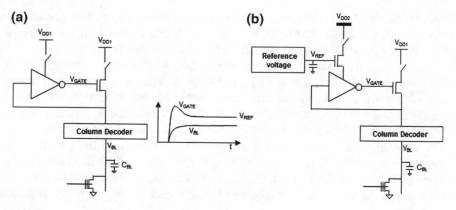

Fig. 4.20 Other bitline precharge schemes (**a** and **b**)

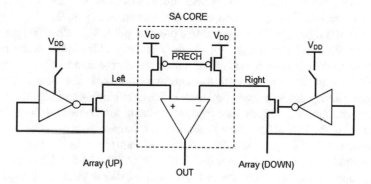

Fig. 4.21 Double-ended sense-amplifier scheme

The inverter stage is supplied through the source of an nMOS follower having the gate supplied by a reference voltage.

This solution allows biasing the inverter with an optimum and controlled voltage level, thus decoupling the structure from the external-supply voltage and enhancing the noise immunity.

Moreover, the supply of the inverter stage takes advantage of a self-boosting mechanism during the pre-charge turn-on transient, thus further increasing the pre-charge efficiency without significant effect the average DC consumption of the inverter stage. Special care must be taken with the dynamic response of the inverter stage in order to avoid underdamped transient, which would delay the bitline settling time and consequently the overall sense-amplifier time response.

Design solutions aiming to control the system open-loop gain for transient-response optimization are disclosed in the literature [20]. A double-ended, symmetrical, and balanced sense amplifier is shown in Fig. 4.21.

The left and right nodes are connected to the drain of the two follower transistors of the bitline pre-charge circuits, thus sinking both the cell current and the reference current. This kind of architecture shows excellent noise immunity, but special care must be taken to avoid differential noise on the sources of the two memory arrays that are usually located on the opposite sides of the sense amplifiers.

Concerning the analysis of the current offset of the current comparator, a dynamic error could be introduced by the circuits of bitline pre-charge if the sensing operation occurs while the current component required to charge the bitline capacitance is still significant with respect to the reference current. It follows that a proper matching of characteristics of the two bitline pre-charge circuits is required. The sense amplifier CORE circuit is also prone to introduce a current offset due to the variation of MOS device characteristics. For this reason, the MOS must have a large gate area in order to minimize the mismatch effect. Regardless, the use of large devices collides with the requirements of speed and dynamic-consumption reduction.

Techniques of offset compensation can be used to limit the impact of unavoidable MOS-mismatch effects and are available in the literature [21]. An example of an offset compensation scheme is shown in Fig. 4.22.

The left and right signals are previously pre-charged to V_{cc}. When the pre-charge is released, the two left and right signals drop linearly with a slope proportional to the current sunk at the two inputs of the bitline pre-charge circuit. Those signals are dynamically transferred to the core through two capacitors. The core consists of a latch preventively set in a metastable state. Specific biasing techniques can guarantee the unstable equilibrium state by compensating for the mismatch of the MOS transistors of the latch. The commutation of the latch into one of the two stable states is driven by the dynamic-differential signal transferred by the two capacitors, which provide the means of decoupling the nodes of the latch with respect to the left and right nodes. This technique allows smaller MOS devices to be used for the latch structure, thus improving the sensing-time response and reducing power consumption. The capacitors can also have reduced dimensions. Another sensing scheme suitable for low-power 1Tr-NOR flash applications is shown in Fig. 4.23 [22].

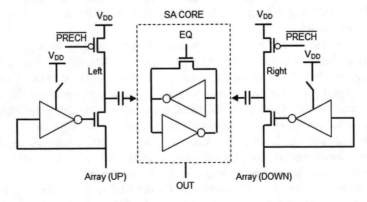

Fig. 4.22 Offset-compensation scheme

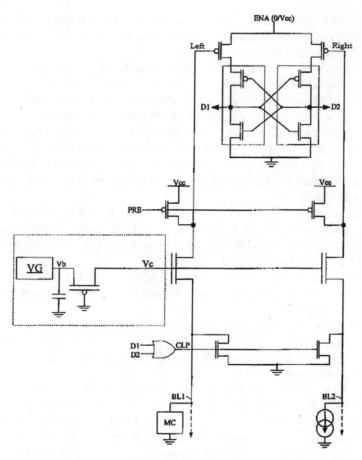

Fig. 4.23 Another sensing-amplifier scheme suitable for low-power applications

The bitline pre-charging scheme is equivalent to the one shown previously in Fig. 4.19b, in which the resistor has been advantageously replaced by a p-channel MOS. The latch stage is equalized during the pre-charge phase. The differential voltage driving the latch commutation is generated between the linearly decreasing signals, both left and right, and it depends on the parasitic capacitance of these two nodes, which are identical due to the symmetrical layout of the circuit. When one of the two signals decreases enough to turn on the corresponding p-channel device, the latch reacts.

The "latch-based" dynamic-sense amplifier, however, requires a controlled signal sequence to set the timing of the different phases of the sensing operation. The relative timings between those signals can be generated using analogue delays, thus accepting the unavoidable timing dispersions due to process drift or temperature and supply voltage variations. Alternatively, it is possible to generate the sensing control-signals sequence through a state machine clocked by a ring oscillator. This solution shows several advantages because a ring oscillator can be independent on supply voltage and temperature variations, and it can be tuned during the device test flow.

This approach requires a re-synchronization of the oscillator to the read-operation command, particularly in the case of interrupted and re-started read operations. Moreover, the ring oscillator must be stopped at the end of the read operation to avoid the current-consumption contribution in idle mode.

Specific design techniques are used to keep the ring oscillator in the off state for a certain time (for example, tens or hundreds of μsec) while keeping the status of the internal nodes at the required working point. This permits a quick restart when requested. A period of long idle modes are covered by periodic "refresh" phases as shown in Fig. 4.24.

STMicroelectronics has been able to achieve dynamic-current consumption of approximately 1–1.5 μA/MHz/bit while granting random-access time within 20 ns in a 40-nm technology node. By trading off with access time, current-consumption values <1 μA/MHz have been demonstrated.

It is important to note that in the definition of the key basic low-power operating modes previously analyzed, the data-modify operations (erase/program) have been deliberately neglected.

Indeed, in a 1Tr-NOR flash, the power consumption associated with those operations is far from negligible, particularly if the relevant power-consumption inefficiency is taken into account. This inefficiency is characteristic of the 1Tr-NOR flash cell programming operation.

The design of the HV generation, regulation, and management circuitry (charge pumps, regulators, level shifters) requires deep design optimization to minimize the unavoidable associated power-consumption inefficiencies. Nevertheless, considering that most of the commercial ultra low-power applications use the embedded memory mainly for pure code-storage purposes, which minimize the demand of data alteration in the typical user-mission profile.

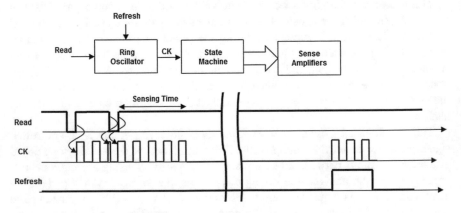

Fig. 4.24 Sensing control-signal sequence

Some of the previously described architectural and design solutions have been successfully implemented in ultra low-power 1Tr-NOR flash macrocells that are embedded in commercial products. Particularly, the high reliability characteristics of the 1Tr-NOR flash cell (which can be further enhanced through ECC implementation) have made it possible to embed ultra low-power 1Tr-NOR flash macrocells in MCUs for implanted medical applications (e.g., pacemakers). STMicroelectronics has conceived and manufactured an ultra low-power 1Tr-NOR flash macrocell in 90-nm technology node that is capable of reaching a typical static-current consumption in Idle mode <10 μA and a read dynamic-current consumption <100 μA/MHz.

4.4 Automotive 1Tr-NOR eFlash Architecture and Its Applications

The evolution of automotive MCUs is driven by multiple requirements, among which are the need for real-time control, the pervasive increase of safety related self-check and diagnostic controls, the high level of reliability, and the capability to update data and adjust parameters multiple times across the entire life cycle of the device.

Complex algorithms, such as those for the reduction of consumption and emission, real-time processing, and dynamical switching between multiple cores executing independent operating functions have sustained the growth of memory performance and size, which is ≤ 8 MB in 40-nm tech node with a forecast for 16 MB for the next node.

Data throughput in read mode is a key factor for MCU performance, particularly considering that everything must be achieved within the full automotive junction temperature range (−40 to 165 °C) until the end of the device's life cycle (20-year data retention) and with tight reliability requirements (<1 ppm). Parameters and data update in automotive systems must target the endurance characteristics of an external EEPROM device, which can ultimately reach 500 K cycles.

In order to stick to existing examples of actual implementations, hereinafter will be described how STMicroelectronics has successfully developed its eNVM-technology concept in the automotive market with in-house 1T-NOR flash cell by exploiting flash cell characteristics to fulfill such tough market requests. In the late 1990s, STMicroelectronics' path in automotive MCUs started with 0.5 μm followed by products at 0.35 μm, 180, 90, 55, and 40 nm in line with the scaling of process technology.

The evolution of code and data flash design and architecture

Automotive requirements motivate the eNVM memory-specifications evolution, which has been addressed at STMicroelectronics by design architectures based on existing 1T-NOR flash process technology. In order to take the maximum advantage from the characteristics of the 1T NOR architecture, the design of code memory and data memory has been branched starting from the common archetype of 1T NOR flash memory existing at 180 nm.

Even if already mentioned in previous sections, it is worth repeating that 1T-NOR flash memory has a quite asymmetric timing behavior in write operation. Programming accounts for approximately 1 μs of physical time, which leads to 20 μs for a 32-bit word under typical process-voltage-temperature (PVT) conditions. The erase operation is divided into three main sub-steps:

- pre-programming
- physical erase
- depletion recovery

Physical erase itself is in the range of 100–300 ms and is almost independent from block size. Pre-programming and depletion-recovery lengths are proportional to block size. Even reducing the block size to the minimum possible, erase time cannot be reduced lower than a certain duration. This is unacceptable for some applications where specific data must be updated in runtime. The main concerns from an application point of view are data availability (need to fetch from flash) and managing long time-out delays.

Monolithic 1T NOR flash arrays are suitable for ROM-like behavior in the case in which content update does not occur while the application is running. In fact, flash technology was originally born as an EPROM replacement adding electrical erase capability. Otherwise, the need to update the NVM content in real time requires the development of a handling technique to overcome the above-mentioned two main concerns associated with an erase operation.

An NVM runtime update, often mentioned as a "parameters or data update," requires a complex algorithm for a new-data entry write, which must have a physical erase because it involves other data entry. These techniques are also referred to in literature as "EEPROM emulation" and use at least two blocks in such a way that new content is first written into an empty block, then other data are then copied, and finally the previous block is erased in order to accommodate the next data write.

Suspend-resume features are a low-cost answer for these systems with low to medium performance, where an NVM update occurs quite rarely and where real-time reaction is not a key element. The suspend command offers the ability to interrupt the write algorithm, thus discharging HV nets, saving relevant conditions, and restoring read bias. The suspend command is characterized by its latency parameter—the time to come back to available read mode—which generally has a maximum value in the range of 10 μs for program and 20 μs for erase. When the suspend sequence has been accomplished, the memory can be read. It is clear that the SW routine to handle the suspend operation must be stored in RAM, which brings additional restrictions. Other limitations to be considered in systems using the suspend feature are as follows:

- the handling of interrupts and service routine, which cannot be kept in flash
- the high suspend-resume rate affects write timing because the charge-discharge of HV nets at each transition, plus the recovery time needed by the algorithm, must be added to the physical time of the suspended and resumed write operation

Read-while-write (RWW) memories are a more sophisticated solution for an NVM runtime update. In RWW architecture, array blocks are organized as partitions that can be independently read or written. Typically, partitions are independent from both a logical and a physical point of view; this means that address and data paths are duplicated as are the HV switches required to handle write voltages. The main reason for that is to limit the area of the devices area involved in HV endurance stress, to reduce the leakage current at high temperature sunk by internal HV supplies (generally internal charge pumps), and to avoid cross-disturbance.

Therefore, from an application perspective, RWW memory is suitable for a clear differentiation path because array partitions are quite well confined from a physical-implementation point of view. Code memory inherits the role of EPROM, as matter of monolithic content that gets written in a well-defined time window in the life of the product, usually when the application is not running. Data memory becomes a flash 1T-NOR array with a high degree of specialization to emulate the word alterability of EEPROM (which, in most of the cases, becomes a replacement in embedded systems of the external EEPROM component).

4.4.1 Code Flash Specification and Architecture

In most automotive applications, code memory is programmed once soldered on the electronic control unit (ECU), during different phases:

- ECU manufacturer plant: Functional test, burn in, and application-code download. Tests are executed at 3T (room, high, and low temperatures).
- Tier 1 (part manufacturer): At tier 1, there is partial or complete reprogramming through a controller-area network (CAN) interface. At this stage, most parts of the application are programmed covering two thirds of the flash space (boot code is excluded). Temperature ranges between room and medium to high temperature (25–100 °C) values.
- OEM (Vehicle Plant): Generally, boot code is excluded in order to allow partial or complete code memory reprogramming and data memory programming (through CAN interface). This operation occurs at room temperature.

Code flash is usually organized into three main sections, i.e., boot, application software, and application data and parameters, as represented in Fig. 4.25. Boot is composed by small- to medium-size blocks, whereas the remaining part of the code memory uses 128- to 512-KB blocks due to area efficiency, at least looking at the last decade's products. Whenever an erase operation is necessary, it becomes the

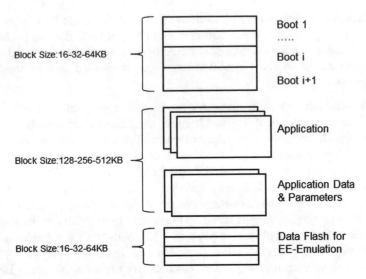

Block Size:16-32-64KB

Boot 1
.....
Boot i

Boot i+1

Block Size:128-256-512KB

Application

Application Data
& Parameters

Block Size:16-32-64KB

Data Flash for
EE-Emulation

Fig. 4.25 Mapping of the use of application blocks

most time-consuming operation; in small blocks it is dominated by the F-N tunnel, with the worst case being at a cold temperature, whereas in large blocks it is dominated by depletion recovery with the worst case being at a hot temperature.

Factory programming

Considering that overall cycle time (program plus erase) in large memory arrays is subject to program and depletion recovery, which share the same physical mechanism, optimizing the modification performance requires squeezing the program times.

It is easier to control voltages within during factory operations than it is in the field. The generation of the drain voltage can be improved by increasing the program parallelism, which plays a major role in the program-time reduction. Particularly, the current capability can be improved by connecting the output of the charge pump to a dedicated 5 V PAD, the voltage value of which must be granted within ±5% (see block diagram in Fig. 4.26). This allows for more than doubling the effective program parallelism, with respect to an embedded charge pump, and reduces the overall erase time ≤ 25% for large blocks.

Erase parallelism

A second technique to improve erase time employs the capability to apply erase pulse voltages in parallel to multiple blocks. Timing improvements are in the range of 10–15% for a medium-sized array (2 MB).

Leakage of HV components, especially at the high temperatures seen in the automotive range, is the limiting factor for the parallelism. Leakage can be sustained by oversizing both the positive and negative HV charge pumps, which often

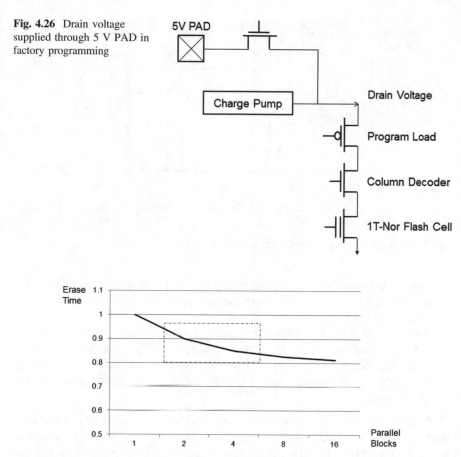

Fig. 4.26 Drain voltage supplied through 5 V PAD in factory programming

Fig. 4.27 Erase-time reduction versus blocks parallelism

results in an area-expensive design. Considering the erase-time trend with respect to block parallelism, the best trade-off can be found by erasing two to four blocks in parallel as shown in Fig. 4.27.

Program buffer

The sequence of operations occurring during a program command in 1T NOR flash has four main contributors, which can be identified with the aid of Fig. 4.28.

- **Fixed**: This contributor is composed of initial and final algorithmic processing, analogue set-up, and analogue discharge and switch off. This part of the sequence is constant regardless of the number of cells to be programmed.
- **Verify**: Pre-pulse verify is performed to determine which cells are to be programmed. Post-pulse verify assesses the effective state of the pulsed cell in case additional pulses are needed.

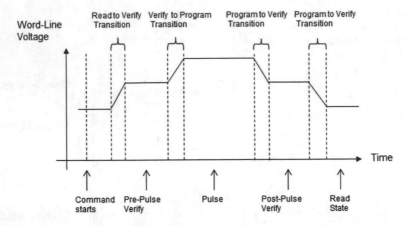

Fig. 4.28 Wordline-voltage evolution during program pulse

- **Pulse**: This contributor is the time required to pulse the selected cells. The duration is modulated by the pulse parallelism and therefore by charge pump-current capability.
- **Analogue transitions**: Word-line and column decoders require dedicated voltage values for each step: read, program verify, and program pulse. Switching these values takes time for analogue set-up, safe switch (free of HV-degrading phenomena such as snap-back and hot carrier injection), and stabilization.

Reduction of the program time can be achieved focusing on one or more of these contributors. Nevertheless, increasing performances through analogue elements, which are designed using HV devices, requires a large investment of area, which becomes less and less scalable with the evolution of the technology nodes. An alternative solution introduces an increasingly larger program buffer, ≤ 1 kb, which maintains the same analogue circuitry and improves the overall program timing. This is achieved by increasing the number of cells involved in each sub-phase so that the weight of fixed and analogue transition sections per bit becomes diluted as can be seen in Fig. 4.29.

Read throughput and line buffer

Access time is a key element in the sizing of high-performance systems, especially real time-operating systems (RTOS), where the response must be as deterministic as possible and within tight time-frame boundaries.

Two elements play a relevant role for random access-time capability: elementary devices used in the decoders and voltage bias. 1T-NOR flash design must deal with the use of HV MOS for row and column decoders, which are needed to sustain high voltages—and which range between 8/12 V for positive and −8/−12 V for negative voltages—in program and erase operations. These decoders are also active parts in read mode, and HV MOS performance is not optimal for high-speed switching. In

Fig. 4.29 Equivalent
cells-programming time

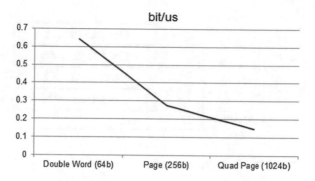

Fig. 4.30 Analogue-read
throughput

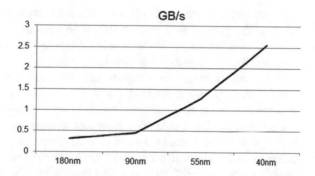

addition, the aging effect of those transistors must be considered, which contributes to the degradation of the end of life access-time performance. The definition of program-verify and EV levels is important for granting program and erase time performance as well as permitting a reliable read window, particularly when read must be performed at very high temperature, e.g., $\leq 165\ °C$ as a junction temperature in the toughest automotive profiles. These voltage values bind the cell-working window to a read wordline voltage of 4–5 V. Considering that logic input and output voltages are driven by low-voltage devices—working in the range of 1.0–1.8 V, level shifters are necessary on the read path, and they contribute an additional timing penalty. By combining all of these disadvantages, 1T NOR flash architecture can produce an acceptable random access-time performance at the high price of relatively large area and current consumptions.

Both of these aspects can be mitigated with wide data parallelism, such as 128- or even 256-bit. Another concept beyond this choice is the use of line buffers, which store read data to be forwarded to the core with minimal latency. In Fig. 4.30, the effective analogue read-throughput evolution is reported for four generations of STMicroelectronics' automotive MCUs.

Instruction and data-line buffers are generally different, and the instruction buffer is the most sophisticated one. Both are managed by a memory controller in order to optimize the use case.

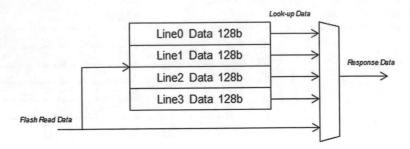

Fig. 4.31 Line-buffer architecture

In medium-performance systems, a memory controller supports a 32-bit data bus width at the port as well as connections to 128-bit read-data interfaces from two kinds of memory array. One flash array is connected to the code flash memory and the other bank is connected to the data memory.

The memory-controller's capabilities vary between the two arrays with each array's functionality optimized for the associated use cases. In Fig. 4.31, for example, the code flash array is managed by a four-entry buffer with each entry containing 128 bits of data. The controller in this case may be configured to allow linear pre-fetch of sequential lines of data from the code flash array into the line buffer as already described in the "Read" section of the "Low-power embedded NOR flash macro-cells" paragraphs of this chapter. In contrast, the controller logic may support only a 128-bit register service as a temporary storage, without any pre-fetching mechanism, when dealing with data flash array.

The pre-fetch buffer hits from the code flash bank support zero-wait data-forwarding responses. Read requests that miss the buffers generate a physical flash read request, which must be managed by multiple wait-states.

4.4.2 Data Flash Specification and Architecture

Data flash is used in conjunction with code flash in a RWW system with the purpose of easing EEPROM emulation. The main characteristics for this element are as follows:

- Write granularity: When dealing with an emulation algorithm, the smallest possible granularity is required, ideally by byte. In case of error-correcting code (ECC) use, the area efficiency of such a choice is unacceptable, and a 32-bit word is generally used as a compromise.
- Read speed is not a key factor because the time latency to access external EEPROM is on the order of µs. The typical access time ranges from 50 to 100 ns with benefits in area and power consumption.

- Program time is generally not critical, and as a reference figure, it takes 20–30 μs for a 32-bit write including ECC. The program-suspend feature is not a particularly interesting option considering the usual 10-μs latency back to read. Bearing in mind the time required for a word program, in most cases the suspend latency equals, on average, the time to complete the operation itself.
- Erase time is a key parameter because the capability to write runtime new data is gated by the HV operation. If an application requires reading data during an erase, either these data must be temporarily stored in RAM for manipulation, or the erase must be suspended. Therefore, the availability of the erase-suspend command is taken for granted. Erase time in 1T NOR flash is directly affected by its endurance, and application time-out is very demanding for erase time as an end-of-life (EOL) parameter.

Erase is generally performed on a single block. From an analogue perspective, this allows fast slopes for charging and discharging HV nets and eases the program erase-controller management as shown in Fig. 4.32. In contrast, the controller must precisely track the algorithm's evolution due to the possible occurrence of a high rate of suspend requests, which may be in the range of 100 μs. In this case, the voltage ladder applied to the cells gets sliced, and the effective cumulated voltage must be saved and restored across a suspend–resume cycle.

Endurance, i.e. the number of cycles that can be executed in the block, is also a key specification element. Data flash must be designed carefully, especially with regard to the sizing of the peripheral circuits, to be able to sustain the cumulated stress while performing the maximum number of cycles on each block. In addition, any bias stress on the array must be limited to avoid any disturbance across different blocks; at the block switch level, any common high-voltage net must be gated.

Typically, the endurance specification for a block is 100 K cycles, which STMicroelectronics automotive MCUs has increased, since the 55-nm generation, to 250 K cycles. The cell's working window must be sized in order to grant EOL performance in the automotive temperature range and especially at high

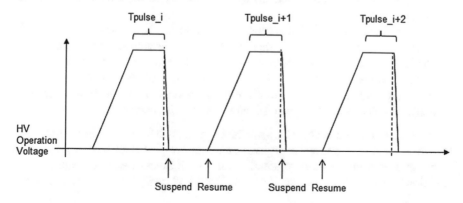

Fig. 4.32 Effect on erase time of high-rate suspend–resume sequence

temperature (165 °C) where 1T NOR flash wear-out phenomena, such as gain and sub-threshold slope degradations, become relevant.

Enlarging the read window has a direct limit in the degradation of program and erase performance. Therefore, sizing requires an optimal mix of architectures and design techniques including specialization of the verify levels for the data memory.

EEPROM emulation with 1T NOR FLASH

The main differences between the embedded 1T NOR flash memory of an MCU and serial external EEPROMs are their access times and write and erase times. Embedded 1T NOR flash has significant advantages with a shorter write access time (20–50 ns vs. µs), which means that critical parameters can be retrieved faster in the emulated EEPROM than in a serial external EEPROM. This speed improves the robustness of the system-reaction path. The emulated EEPROM has a shorter write time, in the range of 10–50 µs/32-bit word, compared with the time needed for word write in a serial external EEPROM. Unlike the external EEPROM, 1T NOR flash must manage the block-erase step, which has no counterpart in EEPROM, and which may last from 200 ms to >1–5 s.

Additional design considerations are needed for the embedded system regarding a possible abrupt interruption in the write operation due to power loss or reset. The external EEPROM requires very low power consumption due to the physics of the write mechanism by F-N tunnel. Therefore, appropriate sizing of decoupling capacitors can secure the complete writing process, which is not possible for an embedded flash. Because the erase process of a block in the flash takes a few seconds, a sudden power loss or asynchronous reset event could interrupt the erase process. Therefore, a robust flash-emulation concept should be considered when designing the management software. This means that in order to design robust flash-based EEPROM emulation software, it is necessary to have a deep understanding of the flash erase process, which is based on three main steps.

- ALL0: The cells are moved from the initial content into a programmed "0" state. This operation usually follows the logical address sequencing. Any interruption during this phase can be identified by the content transition between ALL0 and the initial state.
- Physical erase: The cells in the blocks are progressively turned into a logical "1" state. Any interruption during this phase leads to pseudo-random content.
- Depletion recovery: Over-erased cells are softly programmed in such a way as to remain seen as a logical "1." This is a very sensitive phase because an interruption would leave an apparently erased memory, although the presence of depleted cells would compromise subsequent write operations.

Different emulation concepts are available and can also be found in the literature. However, each concept relies on partitioning an array of flash blocks into a data record and on using pointers to identify valid flash blocks and valid data records. Additional header structures, such as checksum and parity bits, are used to assess

Fig. 4.33 EE-emulation
data-record structure

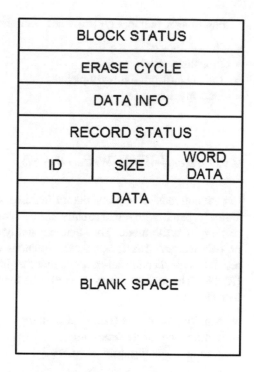

the integrity of the data record. An example of the data-record structure is shown in Fig. 4.33.

The endurance requirement is computed considering the average number of data-record updates with respect to the ratio between block and data-record sizes. When this number is greater than the flash write/erase endurance characteristics, more than two flash blocks are needed to fulfill the requirement, and several techniques are used for prolonging the service life, or wear leveling, in this case. Particularly, safety-system relays, despite any sizing aspect, on three banks that are used alternatively; in the event of a failure (hardware or non-recoverable error) inside a bank, two available banks still remain.

Important considerations must be taken into account when the EEPROM-emulation concept must deal with flash using ECC: First, most of the content left in the memory when an interruption occurs leads to fake ECC-detection flags. As matter of fact, the emulation concept must distinguish between true ECC events due to a data-retention issue, for example, and fake ECC events induced by a write interruption. This is accomplished by a more sophisticated use of pointers and flags. Updating pointers and flags requires an overwrite. This is where the capacity to preserve ECC parity is really helpful, and an ECC-hamming matrix design keeps ECC parity coherent because multiple states must be coded through subsequent writes.

Flash block pointers typically use four states:

- Erased–unused
- Program started
- Data-record program completed
- Data invalid

4.4.3   Read-While-Write (RWW)

In automotive MCUs, a mandatory feature required by the majority of applications is the capability to concurrently access code memory in read mode and data memory in write mode. The same capability may also occur within code memory, or data memory itself, but these situations can be managed by a system work-around. This concurrent-access situation is generally called "read-while-write" (RWW). The main architecture configurations that implement the RWW feature are as follows:

- Multiple memories (i.e., dual memory)
- Independent-arrays memory
- Resource sharing-array memory

Each implementation has advantages and disadvantages that can be summarized as follows:

1. **Area**: Area is reduced by maximizing the sharing of common resources.
2. **Disturb immunity**: Common resources are paths for cross-noise propagations.
3. **Scalability**: Across an MCU family, for a given technology node, the memory size can be increased/reduced as can block composition. Array independency makes it easier to derive embedded 1T NOR flash macrocells with smaller or larger memory cuts.
4. **Development time and cost**: Resource-sharing increases design, layout, and verification effort. In addition, validation activity is affected because disturbance effects can be discovered in a late stage of development, thus affecting yield or customer returns.

Read-While-Write Architecture

In order to allow independent read and write capability, it is necessary to ensure at least two banks of sense amplifiers. A straightforward role is played by the sense-amplifier bank in reading mode, whereas when writing a 1T NOR flash, a read-verify is performed after each type of pulse to assess if the V_{th} (threshold voltage) shift has reached an expected level or if further pulses must be applied.

Both read and read-verify operations require a connection from the bit line to the sense-amplifier bank. This connection is usually hierarchical: block bit lines, also called "local bit lines," are decoded and connected through block column decoders to read main bit lines, which run over the multiple blocks of the array and then reach the sense amplifier.

Write–erase pulses are applied by blocks, and therefore high voltage-management circuitry must be part of the block itself. This is because the voltage bias only involves wordlines and body—a qualifier of the block itself in 1T NOR flash-in erase pulse operations by F-N tunnel. Write-program, using the advantages of physical channel hot electron (CHE), involves wordlines and source lines but also bit lines bias. Indeed, write-program is performed by current switches called "program load," capable of driving ≤ 200 µA, selectively reaching the drain of the cell to be programmed in the selected word. These program loads can be shared across multiple blocks through a "write main bit line," which lay over the array as well. Program bit lines are also used to connect blocks under modification to verify sense-amplifier banks.

Any connection between blocks, sense amplifiers, and program loads requires careful design considerations:

- "Read main bit lines" have length constraints due to capacitance, C, which plays an important role in the access time.
- "Write main bit lines" have length constraints due to IR-drop generated by high current pulse in 1T NOR flash.

Read and write main bit lines have to fit the width pitch of a data bit, and therefore layout spacing between these lines is limited. From an electrical point of view, a key point in the design of a RWW system is the capacitance coupling across main bit lines producing an aggressor-victim eco-system:

- Read to verify (and vice versa)
- Program to read
- Erase to read

Read to verify: In both operations, sense amplifiers are involved. Disturbance may be propagated by power and ground cross-talk. This phenomenon is even more severe because the two operations are asynchronous. Therefore, noise may occur in correspondence with the high sensible data-discrimination phase. Fast read current-peaks disturb precision-sense amplifiers in both read and verify modes. The read operation is directly linked to system performance such as speed and data size. Fast toggling of a large amount of data induces voltage perturbations, which are transferred across the power-ground rails to the verify circuitry.

Verify may disturb read as well. In an embedded system, without input/output (I/Os) toggling, the size of the data involved in read and verify is similar, and therefore it is equivalent to the cross-disturbance. Decoupling techniques are needed to reduce this effect as can be seen in Fig. 4.34

A system that does not allow read-timing overlap is also used in order to avoid transitions during the critical phases: Read timing is bound by access-time constraints, and therefore verify gets stretched with a small penalty (see Fig. 4.35).

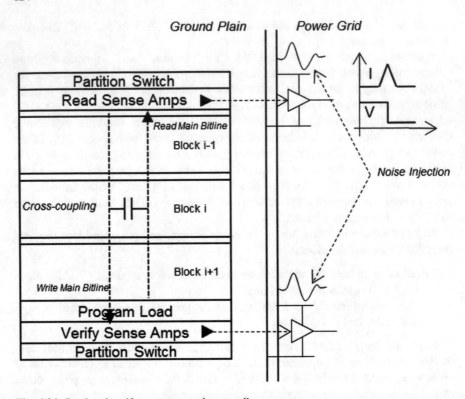

Fig. 4.34 Read and verify power-ground cross-talk

Fig. 4.35 Discrimination
stretch technique

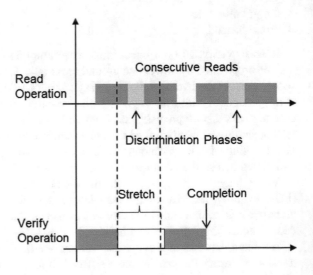

Read and verify may also disturb each other as in the case of coupling across the effect during bit-line pre-charge and data discrimination where the main bit-line voltage gets charged or discharged. The entity of the induced current in the victim node can be evaluated under the terms in Eq. 4.1 where δv is in the range between 0.5 and 1.0 V; δt is approximately 10 ns; and C is approximately 100 fF. Cross-talk current results in 10 μA, which is approximately the same level of signal current to be discriminated into the read and verify modes.

$$di = C\frac{dv}{dt} \qquad (4.1)$$

Equation 4.1: Induced current noise in victim node.

Program to read: Abrupt voltage transition on the program bit line induces charge-sharing into the reading main bit lines. Considerations about this coupling are similar to the ones of the read case. Indeed, δv is in the range of 4.5 V, which amplifies the effects of this disturb. Moreover, it is more complex to implement any kind of dis-overlap because the program-pulse length must be of a precise duration.

Erase to read: There are two main components for this disturb: Wordline to read main bit-lines coupling, and induced coupling of the write main bit line to the read main bit line.

Wordlines of the involved blocks to be erased, which may be multiple, and eventually the whole partition, lay in a horizontal plane and are coupled to read main bit lines, which run vertically in a different metal level. For many layout reasons, the two metal levels may be stacked. The resulting coupling capacitance is very small, e.g., in the range of 0.1 fF, but the multitude of crossing points could exceed a thousand. All of the wordlines are biased at the same level (negative voltage) and therefore can be considered as a single net. The cumulative effect exceeds 100 fF and becomes meaningful during the beginning and the end of the erase pulse. At the start of the negative bias, the slope is quite sharp because the charge pump is able to drive the maximum current. At the end of the pulse, however, the discharger must react quickly. Design solutions are implemented to smooth the negative bias (constant slope), thus helping to reduce the induced electrical disturb. Similar techniques are less effective in the discharge phase because this timing is more critical.

During erase pulse, the positive voltage applied to the body of the block (isolated P_{well}) is transferred by a distributed P–N diode to the drain of the cell itself (N+ diffusion). Therefore, it is transferred to the local bit line, thus producing an effect similar to that of the read-to-verify already discussed.

Layout solutions to contain these coupling effects exploit the benefits of the use of multiple metal levels by creating a grid of grounded shields as shown in Fig. 4.36. Generally, they are effective, but they have the side effect of increasing the overall main bit line capacitive load, which is negative for the read speed.

An additional constraint in the RWW design is the time needed to revert array-voltage bias from the write to the read mode. In the case of program, this time is a significant fraction of the write time, and then it must be minimized. In erase,

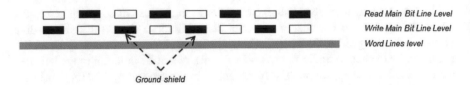

Fig. 4.36 Shielding of main bit lines

the write- to the read-voltage transition time is negligible with respect to the overall erase time. Nevertheless, erase suspend is characterized by a time latency, which is a key parameter, and this means that restoring the read-array bias is a challenge in erase. This bias transition involves several HV nets, among which the most critical are the gate and the source bias. Incomplete or excessive gate discharge would produce a significant disturb in the partition kept in read mode when blocks in write mode are connected back to the gate read-voltage internal regulator. Incomplete source discharge would produce either reliability effects in the source-bias circuitry or high-current peak injection into the ground line with related cross-talk effect. Discharge time is linked to the size of the blocks in the partitions as an effect of the overall capacitive load.

All of the above-mentioned constraints must be taken into account in the design architecture definition of the partitioning scheme. In summary, partitions organized as fully independent arrays are a robust approach, which may result in a less optimized area. Partitions may share periphery resources, such as program load and switches, thus becoming more attractive from a cost perspective but with an increased risk of cross-disturbance.

Stall-While-Write (SWW) Architecture

A low-cost application may not be able to afford the area overhead necessary to implement the circuits required for handling two or more partitions. Stall-while-write makes it possible to initiate hardware emulation with limited resources. The resulting emulation cannot match RWW in terms of performance, but at least it provides a similar software interface. This technique can be applied in the low-end versions of embedded MCU families.

The SWW concept supports various programmable responses for read accesses, whereas the embedded memory is busy performing a write (program or erase operation). The memory controller receives information about the state of the flash array to determine if it is busy while performing some type of write operations, meaning to say that the array is busy.

A dedicated control register defines the response in the case of read access. Typical responses could be as follows:

- Any attempted flash read to a busy array is immediately terminated with an error response, and the read is blocked in the memory controller and not seen by the flash array. Therefore, an error response would generate an exception that must be managed from random-access memory (RAM).

- Basic SWW capability: The memory controller stalls any read master unit until the flash has completed its write operation. If a read access arrives while the array is busy or if the array becomes busy while a read is still in progress, the data phase becomes stalled, and the address and attributes are saved into holding registers. Once the array has completed the write operation, the memory controller reiterates the read request using the saved address and attribute information to create a pseudo-address phase cycle to "retry" the read. Once the retried address phase is complete, the read is processed normally. Once the data are valid, it is forwarded, thus terminating the system-bus transfer. This implementation is fully transparent to software, although watchdog must be carefully managed.
- Enhanced SWW capability uses the ability to suspend any write operation if a read access is initiated. The read request is captured and retried as described for the basic SWW, plus the write operation is suspended by the memory controller. An abort-notification interrupt is generated in order to ensure the appropriate restart.

4.4.4   Typical Example of Implementation at STMicroelectronics

Between 2008 and 2015, STMicroelectronics developed microcontrollers for automotive use with embedded flash featuring data flash and code flash. In Fig. 4.37, the size trend across three technology nodes and the evolution of three architectures of a low-end device with 0.5-MB code memory and 64-KB data memory is reported. Dual flash (two instances with the same organization), optimized data flash (optimized organization of data flash), and full RWW architecture is depicted.

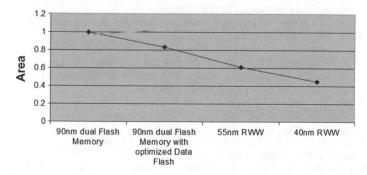

Fig. 4.37 Flash size trend for embedded 0.5-MB code flash and 64-KB data flash

4.4.5 Conclusions

1Tr-NOR flash cell is a solid and valuable solution for embedded flash applications.

It has been on the field since early 1990s, and more than six or seven technology generations have been using this cell architecture with very competitive performances and costs for almost all MCU applications, which are the main motor of embedded flash technologies. In this chapter, we reviewed some of those implementations focusing on secure, low-power, and automotive applications.

There are no technical feasibility concerns at the horizon, at least for the next 2 or 3 technology nodes: stand-alone NVM companies have already been able to put in production in the past years much smaller 1Tr-NOR flash cell sizes and no obstacles are visible with respect to architectural limitations with respect to the major fields of application.

The only concern that is valid for all of the various embedded flash cell structures is this: More cost-effective or better-performing cells will be developed in coming years moving from the early development stage phase, in which they have been in the last decade, to a proven concept based on solid statistical basis. However, so far it seems there is still a long way to go, especially depending on the kind of market field.

References

1. L. Baldi, A. Maurelli, Embedded non volatile memories in deep-submicron CMOS, ESSDERC (1999)
2. A. Maurelli, F. Piazza, Embedded memories. ESSDERC/ESSCIRC, Short Course (2005)
3. S. Marangon, A. Maurelli, M. Moroni, L. Baldi, A salicided flash EEPROM for embedded memory applications. ESSDERC (1996)
4. F. Piazza, P. Colombo, P. Ghezzi, V. Lista, A. Maurelli, E. Palumbo, D. Peschiaroli, S. Soleri, A. Di Biase, A. Silvagni, C. Torti, M. Olivo, L. Baldi, 1.8 μm^2 high density flash memory for 0.35 μm embedded applications. ESSDERC (1999)
5. L. Larcher, P. Pavan, A. Maurelli, Flash memories for SoC: an overview on system constraints and technology issues. IWSOC Tech. Dig. (2005)
6. F. Piazza, C. Boccaccio, S. Bruyere, R. Cea, B. Clark, N. Degors, C. Collins, A. Gandolfo, A. Gilardini, E. Gomiero, PM. Mans, G. Mastracchio, D. Pacelli, N. Planes, J. Simon, M. Weybright, A. Maurelli, High performance flash memory for 65 nm embedded automotive application. 2010 IEEE international memory workshop
7. P. Cappelletti, A. Maurelli, US Patent No 6,410,387 (2002)
8. D. Peschiaroli, A. Maurelli, E. Palumbo, F. Piazza, US Patent No 6,482,698 (2002)
9. P. Cappelletti, A. Maurelli, US Patent No 6,713,347 (2004)
10. A. Maurelli, D. Belot, G. Campardo, SoC and SiP, the Yin and Yang of the Tao for the New Electronic Era. *Proceedings of IEEE*, vol. 97, no. 1 (2009)
11. A. Maurelli, Status and perspectives of embedded non-volatile memories. 2013 International Conference on IC Design & Technology (ICICDT)
12. S. Yamada et al., Degradation mechanism of flash EEPROM programming after program/erase cycles. IEDM, pp. 23–26 (1993)

13. R.S. Scott, R.A. Dumin et al., Properties of high-voltage generated traps in thin silicon oxide. IEEE Trans. Electr. Devices **43**(7), 1133–1143
14. G. Matranga, M. Micciché, R.R. Grasso, Managing of the erasing of operative pages of a flash memory device through service pages. US Patent No 20130272068 (2013)
15. D. Esseni, A. Della Strada, P. Cappelletti, B. Riccó, A new flexible scheme for hot electron programming of non volatile memory cell. IEEE Trans. Electr. Devices **46**(1) (1999)
16. C. Ucciardello, A. Conte, S. Pagano, Charge pump circuit using low voltage transistors. US Patent No 20120250421 (2012)
17. A. Conte, M. Giaquinta, High performance digital to analog converter. US Patent No 20150263758 (2015)
18. M. Giaquinta, A. Conte, R.R. Grasso, F.N. Mammoliti, Background power consumption reduction of electronic devices. US Patent No 8675411 (2014)
19. F. La Rosa, Sense amplifier with fast bitline precharge means. US Patent No 8305815 (2012)
20. F. La Rosa, Bitline bias circuit for non-volatile memory devices. US Patent No 6049491 (2000)
21. G. Campardo, Sense amplifier having capacitively coupled input for offset compensation. US Patent No 5729492 (1998)
22. F. La Rosa, Self-timed low power sense amplifier. US Patent No 8363499 (2013)

Chapter 5
Split-Gate Floating Poly SuperFlash® Memory Technology, Design, and Reliability

Nhan Do, Hieu Van Tran, Alex Kotov and Vipin Tiwari

5.1 Introduction

5.1.1 The Beginnings

More than a quarter of century ago, in summer 1989, NASA's Voyager 2 spacecraft amazed the human world by sending the first image of Neptune and its moon Triton [1]. Neither of those images had anything to do with a flash memory at that time, but coincidently the seeds of flash memory at Silicon Storage Technology, Inc., were being implanted during the same time. Prior to that, in 1984 Fujio Masuoka of Toshiba presented a paper at IEDM describing his flash-memory invention [2], and that inspired various organizations and research institutes around the world to start working on innovative flash-memory technologies.

The primary market driver for flash-memory development was the replacement of ROM and EEPROM technologies for code-storage applications. EEPROM and mask ROM technologies were predominately used for code-storage applications; however, mask ROM, as the name implies, was one-time-programmable during manufacturing, and the contents could not be altered in the field for bug fixes [3]. EEPROM took care of the re-programmability limitations of ROM; however, it was

N. Do · H. Van Tran · A. Kotov · V. Tiwari (✉)
Microchip Technology Inc, Chandler, USA
e-mail: Vipin.Tiwari@microchip.com

N. Do
e-mail: Nhan.Do@microchip.com

H. Van Tran
e-mail: Hieu.Tran@microchip.com

A. Kotov
e-mail: Alex.kotov@microchip.com

© Springer International Publishing AG 2018
H. Hidaka (ed.), *Embedded Flash Memory for Embedded Systems: Technology, Design for Sub-systems, and Innovations*, Integrated Circuits and Systems, DOI 10.1007/978-3-319-55306-1_5

still expensive because the bit-cell size was large, so there was a need for a cost-effective non-volatile memory solution.

Many semiconductor critics at that time were pessimistic about flash-memory technologies because of the perceived inconvenience of erase requirements before any part of the memory could be programmed. However, some innovators were relentless and continued down the flash memory—technology path because of the cost advantages of flash memories for a potentially large market.

5.1.2 Groundbreaking Technology

While the flash-memory trend that was taking place, Bing Yeh and Ching-Chang developed an innovative split-gate flash-memory solution in 1989 and started Silicon Storage Technologies, Inc. (SST) in the heart of Silicon Valley. The technology was named SuperFlash technology because it was perceived to be superior to other flash technology because of efficient and low-power poly-to-poly erase and very efficient source-side injection-program operation. The major differentiator of the SuperFlash technology was due to a specialized and innovative floating-gate tip, which gave it a significant differentiation in the erase operation. Subsequent sections of this chapter will cover the process and design advantages of SuperFlash technology in detail.

Intel was the first one to introduce a stand-alone flash-memory solution to the market with its 1T stacked-gate flash technology, but SST quickly followed with its split-gate based 1-Mb stand-alone flash product (Fig. 5.1), which had a greater acceptance in the market because of its low-power erase and program operations (the details of these functional operations will be covered in a subsequent section of this chapter). Later, when Microsoft announced Windows95, which enabled field updates to bios, the market quickly changed from an EEPROM-based to a flash-based code storage-based solution.

Fig. 5.1 SuperFlash memory wafer from 1996

After its success with Windows95-based platforms, which required a 4-Mb Bios, SST went public in 1995 on the NASDAQ stock exchange (ticker SSTI) and continued to ramp up its NOR flash business. SST's stand-alone NOR flash business peaked at $450 million in 2004. SST thereafter encountered some headwinds in the market because many other stand-alone NOR flash players emerged in the market and caused flash-memory price erosions. SST's stand-alone NOR flash business continued to encounter challenges in the coming years; however, it was just the beginning because its success in the SuperFlash technology–licensing business was yet to come.

5.1.3   IP Licensing Business

SST's success in the NOR flash-memory market is often overshadowed by its market dominance in the embedded SuperFlash technology–licensing business, but the roots of SST's licensing business are based on the initial success of its NOR flash business. From the beginning, SST understood the importance of intellectual-property protection, and therefore it continued to focus on strong innovation cycles and building a strong intellectual-property (IP) portfolio. SST's differentiated NOR flash solution, along with its strong IP portfolio, later paved the way for multiple foundries and IDMs to adopt SST's SuperFlash technology for a series of embedded flash-based devices such as general-purpose microcontrollers, automotive microcontrollers, smart cards, and various other products.

Today, IP strategy is an essential part of SST's business plan. At the time this book was written, SST had >650 international patents issued and pending. SST has worked very closely with worldwide foundries and IDMs for transferring technology know-how in the semiconductor-factory and -design communities, and these close collaborations have resulted in building a strong ecosystem of SuperFlash technology-enabled products.

5.1.4   Chronology of SuperFlash-Memory Technologies

SuperFlash technology, design, and reliability aspects will be covered in subsequent sections of this chapter, but it is worthwhile to address the chronology of various SuperFlash technologies. SST developed three generations of SuperFlash technology termed ESF1, ESF2, and ESF3. ESF stands for embedded SuperFlash, and the suffix (1, 2, and 3) stands for generations 1, 2, and 3, respectively. ESF1 technology was invented in 1989, continued to be the workhorse of various stand-alone NOR and embedded products for several years, and was quite scalable ≤ 180 nm. After 180 nm, ESF1 was losing steam for cell shrink, so SST invented ESF2 technology, which used a self-aligned process for aggressive cell shrink. ESF2 technology was very cost-effective, but it took a long time to ramp up to production and was not a desired technology to scale further. In order to continue scaling SuperFlash technology, SST acquired privately held Actran Systems in April 2005 [4], and that was the genesis of ESF3 technology.

5.1.5 Current Business Overview

Microchip Technology Inc. (Microchip) acquired SST in April 2010 [5], and now SST is a wholly owned subsidiary of Microchip. SST's previous stand-alone NOR flash business is being managed by Microchip while SST focuses on licensing its SuperFlash memory technology and know-how to IDMs and foundries for embedded applications. As of fall 2016, SST has deployed SuperFlash technology at >25 IDMs and foundry partners with >40 SuperFlash technology installations across a series of technology nodes and geographies.

Many IDMs and foundries across the globe have adopted split-gate SuperFlash technology for a series of embedded applications including microcontrollers, smart cards, *Bluetooth*®, Wi-Fi®, ZigBee®, CPLDs, power management and other flash enabled devices. As the cost of non-volatile memory continues to decrease, several other applications exist that can use embedded flash for adding various new features to their existing devices. Figure 5.2 shows the overall shipments and annual shipment rate for SuperFlash technology-enabled devices by SST and its licensees through 2015. As automotive applications and the Internet of Things (IoT) continues to drive the growth of the semiconductor industry, we expect the annual run rate of SuperFlash technology-enabled devices to continue to increase, and the shipment of 100 billion SuperFlash technology-enabled devices is in the line of sight.

Fig. 5.2 Shipment of SuperFlash technology–enabled devices

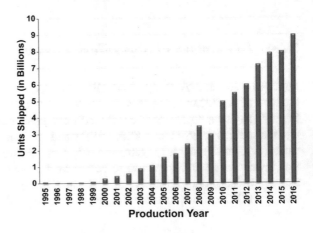

5.2   Split-Gate SuperFlash Technology

5.2.1   Fundamentals of SuperFlash Technologies

The SuperFlash memory cell was invented by Bing Yeh, a co-founder of SST, in 1989 [6]. The memory cell has evolved beyond the initial concept, enhancing the scaling of the cell from a 1-μm technology node to a 28-nm technology node and possibly smaller geometry. Through scaling and evolution, the memory cell retains its split-gate structure, poly-to-poly Fowler–Nordheim (FN) tunneling erase, and source-side hot-channel electron (SS HCE) injection programming [7–12].

The first-generation SuperFlash (ESF1) cell has been used in both stand-alone and embedded flash-memory products for >20 years. Figure 5.3 illustrates the cross-section of an ESF1 cell. The non-self aligned double-poly cell has floating gate poly as a storage element and three terminals for read, erase, and program operations: select gate (WL), source (SL) and drain (BL). The floating-gate and select-gate channels are split between source and drain, hence preventing any over-erase, a characteristic normally encountered in a stack-gate flash-memory cell.

The beauty of the first-generation SuperFlash technology lies in the simplicity of forming the memory cell in a given logic process. In a typical ESF1 eFlash process, only six to seven additional masking steps are required to integrate the memory cell and high-voltage devices for cell operation to the baseline process. The integration is done in such a way that it maintains the baseline device's performance. Due to its maturity and the continuing demand for eFlash in various applications, we have recently integrated this technology in a variety of process platforms—such as high voltage, BCD, TDDI, and more—with the number of additional masking steps being reduced to two to four. These recent developments offer very cost-effective embedded NVM solutions.

The field-enhanced tunneling injector on the floating poly is formed using standard CMOS oxidation and dry-etching processes. The WL poly is inherently the logic poly. The drain is created with logic-junction formation, and the source is a high-voltage (HV) junction that can support programming voltage (9–10 V). The first-generation cell has been scaled from a 1 to a 0.11-μm technology node and will

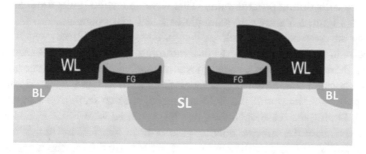

Fig. 5.3 Illustration of ESF1 cell cross-section

likely be adopted for smaller geometry. The memory array is arranged in cross-point architecture with rows of word lines and columns of bit lines. The rows can be segmented into big blocks or small sectors and thus are suitable for a wide range of embedded flash applications.

A highly efficient programming with split-gate flash-memory cell had been discussed in the literature. One of the early works, published by Kamiya [13], described an EPROM cell called "perpendicularly accelerating channel-injection MOS" (PACMOS). In that cell, both the select gate and the floating gate, which are arranged in series between the drain and the source, are biased to form channels under the gates. The floating-gate voltage is formed with capacitive coupling from the select gate and the drain during programming. The channel-injection uses a channel potential gap built at the boundary of these two channels. Similarly, Hu and his team from the University of California at Berkeley published in 1986 [14] about a split-gate cell called "source-side injection EPROM" (SIEPROM) with enhanced programming efficiency by adding a coupling gate on top of the floating gate and a floating poly spacer next to the poly stack. They reported a cell structure that had several advantages over the conventional EPROM cell because the cell could be programmed with a low drain voltage of 5 V, and—even with the low programming voltage—the programming speed of the cell was much faster than that of existing conventional drain-side injection EPROMs. The mechanism of the enhanced hot-electron injection in split-gate transistors was thoroughly analyzed and demonstrated to be useful for EEPROM applications by IMEC researches in 1992 [24].

SuperFlash split-gate structure offers very efficient programming. With a small-channel current for programming, the WL can be slightly biased to turn on the select gate transistor. With high voltage applied to the SL, a strong lateral electric field induces near the FG–WL gap area that generates channel hot electrons. A strong vertical electric field by the floating gate then helps to inject these hot electrons into the floating gate efficiently. Figure 5.4 shows the simulated lateral and vertical electric fields along the channel.

Due to the lack of a dedicated coupling gate for programming in the ESF1 cell structure, a strong vertical electric field primarily relies on the coupling of SL voltage to the floating gate. To optimize this SL–FG coupling, a large source junction under the FG is required. However, an effective floating-gate channel length must be properly scaled to avoid an unintentional disturbance for the un-selected cells during programming. This adverse feature limits the scaling of the ESF1 cell. Figure 5.5 shows an example of ESF1 cell operation.

During an erase operation, voltage is applied to the word lines, which erases all of the cells in one pulse. Similar to programming, verification is not required after an erase. During wafer sorts, certain tests are included to ensure that all bits are properly screened to warrant proper operations until the end of the device's life. Poly-to-poly tunneling with the field-enhanced injector facilitates fast erase time (~ 1 ms). During the erase operation, the WL is biased with ~ 12 V, and electrons are removed from the storage floating-gate poly to the WL by tunneling mechanism. As shown in Fig. 5.6, the contours of the electric field in the tunneling oxide indicate that the sharp tip of the floating-gate poly enhances the electron tunneling.

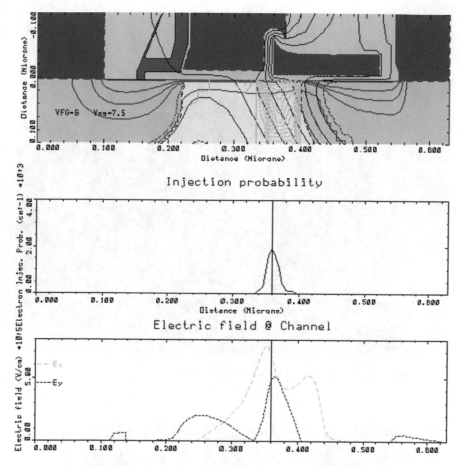

Fig. 5.4 Simulated e-fields during program: **a** Simulated cross-section with potential contours. **b** Simulated location with highest hot electron-injection probability. **c** Lateral and vertical electric fields along the channel

	Vwl		Vbl		Vsl	
	Sel.	Unsel.	Sel.	Unsel.	Sel.	Unsel.
Erase	11-12V	0V	0V	0V	0V	0V
Read	1.8 – 2.5V	0V	0.8V	0V	0V	0V
Program	~1.5V	0V	~ 5uA	~Vcc	~9.5V	0V

Fig. 5.5 ESF1-cell operation

This therefore allows the use of thick tunneling oxide, a feature that completely eliminates the stress-induced leakage current (SILC).

To enhance scaling, the SuperFlash cell was further evolved with the second-generation (ESF2) [15–17] self-aligned memory technology. Although a flash macro with ≤ 4 Mb of memory flash density is ideal for the first-generation

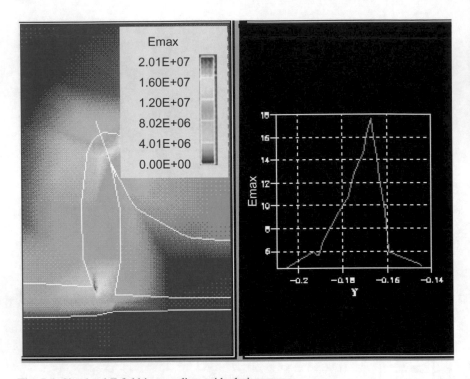

Fig. 5.6 Simulated E-field in tunneling oxide during erase

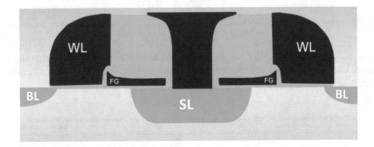

Fig. 5.7 Illustration of ESF2-cell cross-section

SuperFlash technology, the smaller cell of the second-generation cell allows system designers to incorporate an eFlash macro with a memory-array density >4 Mb.

The advancement of the ESF2 cell comes with self-aligning processing steps in forming the floating gate, source line poly, and WL poly, thus reducing the number of masks required to create the memory cell. The floating-gate element is self-aligned to the formation of active in the cell-width direction (x) and to an oxide spacer in the cell-length direction (y). The source-line poly and WL poly are self-aligned to the floating-gate poly and dielectric spacers. As shown in Fig. 5.7, the ESF2 cell retains the fundamental features from the previous generation's cell: split-gate architecture, poly-to-poly FN tunneling erase, and SS HCE injection programming.

With self-aligning processing steps, an ESF2 cell can be significantly scaled. For the same technology node of 180 nm, ESF2 can be scaled >40% smaller than ESF1. Figure 5.8 shows the improvement of ESF2 scaling in the x-direction. FG poly is self-aligned to active formation; therefore, poly overlapping with active is no longer required to account for process alignment and critical dimension control. With the addition of SL poly, the reduction of programming voltage from >9.5 to <7.5 V allows the FG channel–length scaling of >30%. Furthermore, ESF2 decouples the WL oxide, also the tunneling oxide, from high-voltage gate oxide. This decoupling allows the scaling of both WL oxide and channel length. A thinner tunneling oxide not only benefits the WL channel scaling but also improves the cell program/erase (P/E) cycling endurance significantly. Figure 5.9 summarizes the study of endurance as a function of tunneling oxide thickness [18].

The second-generation SuperFlash cell retains the simple operation and array architecture of the first-generation technology. It also retains the performance and reliability of the first-generation technology. The memory cell can be programmed in <1 μ and erased in 1 ms. The erased-cell current is maintained between 35 and

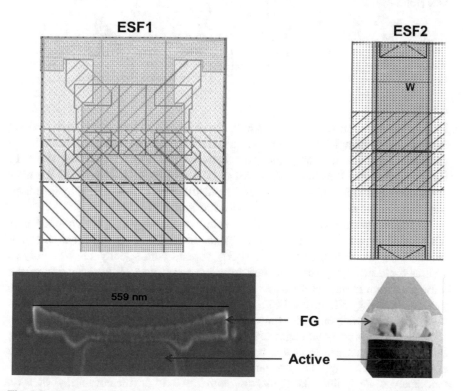

Fig. 5.8 Illustration of scaling improvement from ESF1 to ESF2 in the x-direction with **a** layout sketches and **b** cross-section (not to scale)

Fig. 5.9 Endurance as a
function of tunneling-oxide
thickness: **a** $T_{ox} > 20$ nm for
ESF1 0.33 µm.
b $T_{ox} < 20$ nm for ESF1
0.25 µm. **c** $T_{ox} \sim 15$ nm for
ESF2 $0.18 \leq$ µm. **d** Under
typical application conditions
for ESF2 0.18 µm [18]

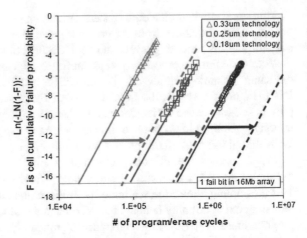

Fig. 5.10 Normal
distribution plot for erased
cell current of 32-Kb array
subjected to 3 million P/E
cycles and 1500 h of 150 °C
bake [18]

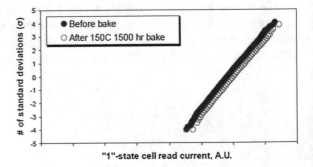

40 µA. The endurance is as high as millions of erase and program cycles, and the
memory cell is completely immune to stress-induced leakage current (SILC) due to
the relatively thick tunneling dielectric between the floating gate and the WL.
Figure 5.10 shows the immunity to SILC of the ESF2 cell array of 32 Kb after
being subjected to 3 million P/E cycles. The distribution of cell current in the erased
state shows no evidence of leakage after 1500 h of 150 °C bake.

To further enhance the scaling of the split-gate memory cell, the coupling gate
(CG) is introduced in SuperFlash's third-generation technology (ESF3) as shown in
Fig. 5.11. In addition, the erase gate (EG)—formed with select gate (WL) poly as
well as logic gate poly—is separated from the select gate. This separation of the
erase gate from the select transistor allows the scaling of the select transistor. The
gate oxide under WL in the ESF3 cell has the flexibility of adopting either the
core-logic gate oxide or the IO-transistor gate oxide; therefore, the WL channel can
be scaled appropriately and aggressively. With the assistance of the coupling gate,
the source junction can be scaled back, thus providing further improvement in the
scaling of the floating gate channel. In short, the aggressive scaling of the ESF3
cell enables designing flash macros with very high memory-array density.

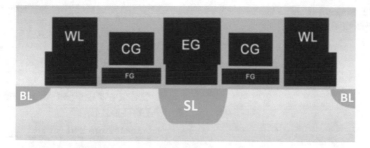

Fig. 5.11 Illustration of ESF3-cell cross-section

This technology has now become the workhorse for embedded flash applications in modern electronics.

The process of forming the third-generation cell is simpler than that of the second-generation cell despite the additional nodes to enhance scaling and performance. The floating gate is self-aligned to the active and coupling gate. Both the word line and erase gate come from the logic poly-silicon deposition. Drain and source junctions come from a standard-logic process. Unlike the sharp-tip requirement for the FG poly in previous SuperFlash cells, ESF3 erase relies on the tunneling between the FG corner and the EG through a thinner dielectric. In addition to the improvement of erase performance, the thinner tunneling oxide also improves endurance as shown in Fig. 5.9.

The addition of the coupling gate and the separation of the erase gate allows the competitive scaling of the SuperFlash memory cell from a 110 to a 28-nm technology node.

Memory-cell operation is fundamentally unchanged from previous SuperFlash generations. The cell is inherently programmed using the highly efficient SS CHE - injection mechanism. With additional coupling from the coupling gate, the source voltage used for programming is significantly reduced, thus alleviating programming disturbances. Flexibility in coupling-gate decoding also helps to expand the program-disturb window, which is required especially in high-temperature automotive microcontrollers. Fast erase is done with poly-to-poly tunneling using the erase gate, which helps to further improve endurance and to maintain the SILC-immunity characteristics.

With the word line decoupled from the high-voltage erase, the select transistor can be optimized for high cell current with core logic-supply voltage. This is ideal for low-voltage, low-power, and high-performance applications.

5.2.2 Cell-Process Architectures and Operations

Despite changes of the cell structure through architectural evolution, the fundamental operation conditions of the split-gate SuperFlash cell have remained

Fig. 5.12 Simplified
ESF1-process steps

| Active |
| HV transistors Well & Vt |
| Memory Cell |
| CMOS Well & Vt |
| Gate Oxide |
| Gate Poly Deposition and Formation |
| LDD and D/S |
| Salicidation and Metalization |

unchanged through several generations of technology scaling. This sub-section describes the process architecture of SuperFlash-cell technology with an emphasis on ESF1 and ESF3, which are better known among eFlash users, and further explores the electrical operation of the cells and the merits behind them.

As shown in the illustrated cross-section (Fig. 5.3), the ESF1 select gate is spaced from the floating gate with a dielectric gap of approximately 15–25 nm. This gap oxide is shared with tunneling as well as the WL and the HV gates and therefore simplifies the integration between the WL and the high-voltage transistors. The thickness of the gate dielectric is optimized for HV operations and endurance as discussed in the previous subsection. To minimize the impact of flash integration to the baseline devices, the process steps dedicated to the formation of the memory-cell and HV devices precede those of baseline logic devices as shown in Fig. 5.12. After shallow trench–isolation steps are the ion implantations to define wells and threshold-voltage adjustments for HV devices. The memory cell-process section begins with the FG transistor, which includes gate oxide, polysilicon, and ion implantations. The floating-gate oxide thickness is flexible depending on the performance and reliability requirements of particular applications. A thick oxide between substrate and FG poly degrades the programming-injection efficiency but increases the immunity to charge loss from FG to substrate or SL junction it is ideally targeted for automotive devices that require near zero failure. The size and shape of the FG is realized with a LOCOS-oxidation step. The thickness of FG poly is optimized for capacitive coupling between the WL and the FG sidewall without a breakage in the middle of FG poly, as shown in Fig. 5.13, which can potentially be a reliability concern.

After the formation of the FG is the WL and HV gate oxide and poly where the logic-gate module will become a part of it. It is worth mentioning that care must be taken during this process module to alleviate the reverse tunneling characteristics between the WL and the FG. As discussed in [19], a low forward-tunneling voltage is needed for fast erase, but a low reverse-tunneling voltage between FG and WL poses a concern for program disturb on the unselected cells. Figure 5.14 shows both forward- and reverse-tunneling characteristics of the ESF1 cell.

The back-end-of-line (BEOL) process of ESF1 integration adopts the available baseline process. One of the advantages in eFlash adopting SuperFlash technology is the minimal impact of the additional processing steps on the logic devices, and

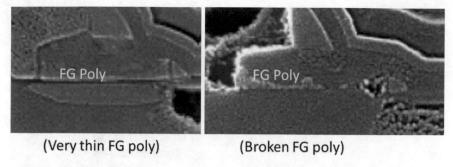

(Very thin FG poly)   (Broken FG poly)

Fig. 5.13 Breakage of FG poly due to thin poly and oxidation

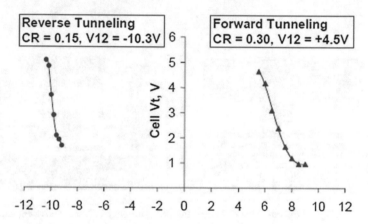

Fig. 5.14 Reverse- and forward-tunneling characteristics [19]

this advantage continues through the third-generation (ESF3) cell technology. Figure 5.15 simplifies the ESF3-process flow where both HV and memory-cell devices are mostly formed before logic devices. This modular architecture not only reduces the impact on logic but also makes the same process node with different "flavors" interchangeable. For example, the ESF3 technology can initially be integrated into a 40-nm low-power (LP) platform and then modularly converted to a 40-nm ultra LP (uLP) without any additional work to optimize the flash devices.

During read, programming, and erase, the ESF1 FG memory cell relies on the capacitive coupling between the FG and the adjacent electrodes. The potential of the FG is governed by:

$$v_{fg} = \frac{Q_{fg}C_{wl} \times V_{wl} + C_{vs} \times V_{vss}}{C_{tot}}$$

where Q_{fg} is the charge in the FG; and C_{wl}/C_{tot} and C_{vss}/C_{tot} are the coupling ratio between the WL and the FG and between the SL and the FG, respectively.

Fig. 5.15 Simplified
ESF3-process flow

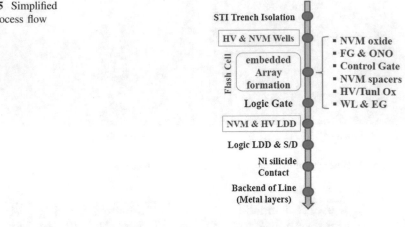

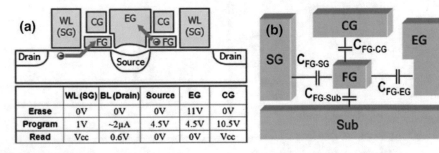

(a)

	WL (SG)	BL (Drain)	Source	EG	CG
Erase	0V	0V	0V	11V	0V
Program	1V	~2μA	4.5V	4.5V	10.5V
Read	Vcc	0.6V	0V	0V	Vcc

Fig. 5.16 The typical operation conditions of the ESF3-memory-cell and equivalent-capacitance diagram [18]

A similar equation, with the additional electrodes of CG and EG, governs the potential of the FG in the ESF3 cell. As found in [20–23], ESF3-cell operation takes advantage of the additional nodes. Figure 5.16 lists the operation conditions and the equivalent-capacitance diagram.

As illustrated in Fig. 5.5, the ESF1 cell is read by applying reference voltages to the select gate by way of the word line and to the drain by way of the bit line while the source is grounded. Reference voltage applied on the word line turns on the select gate portion of the channel; the cell conducts current if the floating gate is erased (low-threshold state). However, the cell is non-conductive if the floating gate is programmed (high-threshold state). The conductive cell outputs logic "1," and the non-conductive cell outputs logic "0." The cell read through the control of the select transistor eliminates the "over-erase" issue encountered in the stacked-gate flash memory cell. Figure 5.17 shows an example of the cell current distribution in both the programmed and erased states.

As previously mentioned, a SuperFlash cell is programmed using a highly efficient SS CHE-injection mechanism. During programming, a high voltage applied on the source, which capacitively couples to the floating gate through

Fig. 5.17 Cell-current distribution in both the programmed (0) and erased (1) states

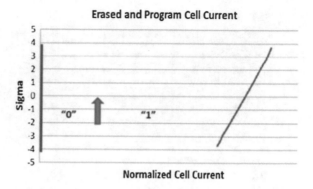

floating gate oxide to create a vertical electric field between the floating gate and channel. At the same time, a voltage slightly higher than the select-transistor threshold is applied to the word line. The extension of the source and the drain voltages toward the thin gap between the floating and select gates creates a strong lateral electric field across the gap, which generates channel hot electrons. The vertical electric field then sweeps these electrons to the floating gate, thus reducing the floating-gate potential to shut off the floating-gate portion of the channel. Such a high rate of electron injection into the floating gate at the gap region of the channel enables the cell programming to be completed in a very short time (~ 1 μs) with very low programming current (1–5 μA) per bit.

Programming performance improves in the ESF3 cell with the addition of the coupling gate (CG) and the separation of the erase gate (EG) from the select gate (WL). Together with the EG and the SL, the CG enables an HV coupling to the FG through an oxide–nitride–oxide (ONO) layer to provide a large vertical electric field during programming. Thus, both the SL junction and the voltage can be scaled down. Furthermore, the FG channel is inverted during programming, thus inducing a strong lateral electric field near the WL–FG gap. With such efficient generation of channel hot electrons, a lower programming current ($\sim \leq 1$ μA/bit) is required to sufficiently inject the electrons into the FG. A reduction of SL voltage during programming also improves the programming disturb immunity, thus making ESF3 ideal for high-temperature operation. Figure 5.18 shows the programming efficiency and program-disturb operation window in ESF1 and ESF3, respectively.

The SuperFlash cell erases using poly-to-poly FN tunneling. With the floating-gate tip (or corner) functioning as a field-enhanced tunneling injector, a lower voltage, compared with a stacked-gate flash cell, can be used on the word line (or the EG) poly to erase the cell. Due to the low coupling ratio between the word-line (or the EG) poly and the floating-gate poly, a significant voltage drop exists across the tunneling oxide. A high electric field is localized along the edge of the tunneling injector creating a very fast charge transfer between the word-line (or EG) poly and the floating-gate poly, and the charge transfer eventually slows down when enough positive charges accumulate on the floating gate. This accumulation

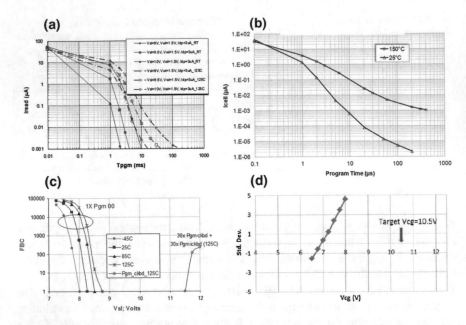

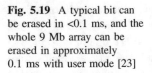

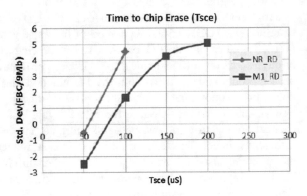

Fig. 5.18 Programming efficiency in **a** ESF1, **b** ESF3 as well as the program-disturb window in **c** ESF1 and **d** ESF3

Fig. 5.19 A typical bit can be erased in <0.1 ms, and the whole 9 Mb array can be erased in approximately 0.1 ms with user mode [23]

of the positive charge on the floating gate increases the floating potential and determines the erased state "1" of the memory cell.

In an ESF3 cell, tunneling oxide between the EG and the FG is decoupled from the WL and the HV gates; therefore, its thickness can be optimized for forward tunneling to enhance fast erase and to improve P/E endurance. Compared with the 1-ms erase time in ESF1 with an equivalent erase high voltage, Fig. 5.19 shows that a typical bit in a 9-Mb array can be erased in <0.1 ms. In addition, a typical array can demonstrate an intrinsic endurance of close to 1 million P/E cycles as shown in Fig. 5.20.

Fig. 5.20 Intrinsic endurance in ESF3 demonstrates 1 M P/E cycles [23]

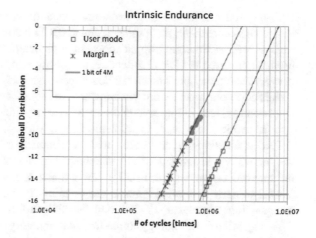

5.2.3   Conclusion

SuperFlash technology has evolved during the last 27 years since its inception into a flash technology of choice for various embedded applications. Despite challenges in integration, baseline-process materials, and scaling, each new generation of SuperFlash technology performs better than the previous generation while retaining its excellence in reliability, which is highly demanded in modern electronics. In addition to embedded applications for automotive in consumer electronics, SST's SuperFlash cell has become a popular choice for smart-card applications due to its fast write time (a few milliseconds) and access time, smaller sector architecture, low power capability, and million-cycle endurance.

The ease of integrating the high-performance SuperFlash memory cell into baseline CMOS logic process has made SST a strong and long lasting partner with several of the world's leading foundries and IDMs, thus allowing SST's SuperFlash technology to be an attractive and dependable solution for embedded flash-system designers.

5.3   SuperFlash Memory-Design Techniques

The trend of functional integration on system-on-chip devices is accelerating due to more transistors per chip being available for advanced node and more user-required functions to satisfy new and broader market needs such as mobile, IoT, security, and automobile controls. Embedded non-volatile memory at advanced node is needed to provide high-density code as well as data-storage applications at reduced cost and with better performance and reliability. The embedded SuperFlash design [44, 45] is to provide flexible IO widths from 8 to 256 bits, varied density, simple macro interface, extremely low power, high performance, area efficiency, reliability, and multi-megabit memory. These are requirements often seen for embedded applications such as consumer, power management (BCD platform), automotive, and smart cards as well as emerging applications such as neural networks for deep learning through optimized flexible architecture, power-aware design, and built-in embedded memory-testing engine. The embedded SuperFlash design for testing and reliability is realized to optimize macro performance, to reduce test cost, to provide superior data retention, and to improve endurance. Memory architectures are optimized for specific applications such as implementing a very small sector size to emulate EEPROM functionality and segmented architecture for low power and high speed. The innovative design has been implemented or demonstrated successfully across multiple process nodes from 350 to 2× nm and at multiple foundries and IDMs. Current designs and developments are ongoing to extend SuperFlash technology into sub-20 nm in planar, FDSOI, and FinFET platforms.

A SuperFlash technology three-terminal memory cross-section is shown in Fig. 5.3. This is the first generation of proprietary split-gate SuperFlash technology with simple integration of a non-self aligned process. This memory-cell technology allows for low mask count because there are only three terminals. This ESF1 technology uses poly-to-poly Fowler–Nordheim tip-enhanced tunneling for erasing and source-side channel hot-electron (SSCHE) injection for programming. Read operation uses logic V_{dd} voltage on its word line (WL).

A second-generation SuperFlash memory, ESF2, uses self-aligning process steps and double poly with three terminals, which is same as that of ESF1 as shown in Fig. 5.7. The self-alignment of the FG to active and the SL and the WL poly to the FG allows for very small cell size comparable with that of the 1T stack-gate cell. Erase, program, and read operations are similar to those of ESF1.

A third-generation split-gate SuperFlash (ESF3), five-terminal memory cross-section is shown in Fig. 5.11. This is the latest generation of SuperFlash technology, which allows scaling of the SuperFlash memory technology into the sub-2× nm regime with self-aligned floating gate (FG) CMP (chemical mechanical polishing), three-poly (WL/logic/EG, FG, CG) thin word line oxide and separate erase gate and control (or coupling) gate. This ESF3 technology uses poly-to-poly Fowler–Nordheim tip-enhanced tunneling for erasing and source-side channel hot-electron injection with control-gate coupling for programming. This memory-cell technology allows reading at logic voltage (V_{dd}), hence providing state-of-the-art read power consumption and area.

The content contained in the SuperFlash memory design section is as follows:

5.3.1 Chip architecture
5.3.2 Chip operation
5.3.3 Array architecture
5.3.4 Decoding
5.3.5 Sensing
5.3.6 HV generation
5.3.7 Design for test and reliability.

5.3.1 Chip Architecture

An embedded SuperFlash chip architecture is shown in Fig. 5.21 that includes essential circuit blocks for implementing all important memory functions and testing methods necessary to ensure data integrity, high endurance (more than hundreds of thousand cycles), and high reliability of the SuperFlash memory technology across multiple power-supply ranges (0.8–5 V) and temperature specifications (−40 to 150 °C ambient temperature). The SuperFlash memory architecture provides small-sector (256–8 K Bytes), fast erase (normally a few milliseconds), and very low-power fast program (approximately 1 μA of programming current in a few microseconds) operation. The IO width of data-in and data-out can range from 8 to 256 bits. The chip architecture typically includes error-correction code (ECC) for high-performance or -endurance requirements. An example pictograph of a 16-Mb ESF3-based memory solution at 40-nm technology is shown in Fig. 5.22.

A logic control-block processes the incoming control, data-in, and address pins to decode the command function such as read, program, erase, or test. The row decoder (XDEC) decodes the input addresses to provide row enabling in read mode. The row decoder also allows multiple rows to be enabled depending on the mode such as sector, block, chip erase, or mass modes. The column decoder (YDEC) provides bit-line selection and de-selection. It also provides a bit line-inhibit function in program for unselected memory cells. The high-voltage decoder (HVDEC) provides an appropriate bias level, high voltage, low voltage, or current

Fig. 5.21 A SuperFlash chip block diagram

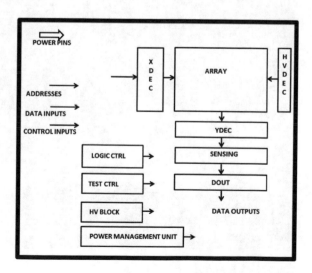

bias for erase gate, control gate, and source line in erase, program, read, and stand-by operation.

The sensing block provides sensing of the selected memory cells by current-mode sensing (meaning current comparison) or voltage-mode sensing (meaning voltage comparison) to translate the analog floating-gate storage information in the memory cell into a logic-level data-out. Different sensing schemes— such as single-ended sensing, differential dynamic sensing, and pipeline sensing— are used depending on performance, power, and area (PPA) requirements for particular applications such as low power, high speed, smart card, and serial high-data bandwidth (>100-MHz clocking) transfer rate. A data-out block (DOUT) provides latching and buffering of the output data from the sense amplifiers.

The high-voltage block generates power-efficient regulated high-voltage levels to be coupled to the selected array terminals in program and erase operation. It includes high-voltage charge pumps, high-voltage regulation, and a sequential program and erase timing-control circuit.

The testing-control block (TEST CTRL) provides various embedded testing functions such as multiple test modes for testing the integrity of the memory array, embedded trimming algorithms for bias or timing, speed-test function, etc. An on-chip power-management unit controls power sequencing, power-on reset, and various power states of the macro such as deep power-down mode, stand-by modes, and active mode to optimize overall system power consumption.

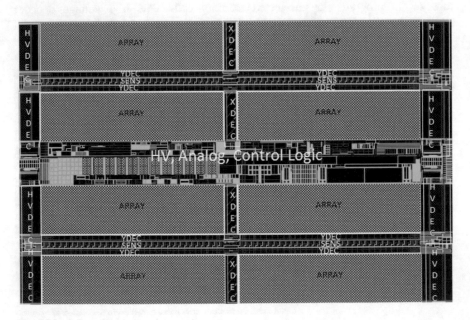

Fig. 5.22 A SuperFlash die pictograph

5.3.2 Chip Operation

The embedded SuperFlash technology design provides all macro functionality needed for the system-on-chip to manage non-volatile memory functions. The macro pin interface includes asynchronous (without CLK) or synchronous (with CLK) interface. Control pins include CE/, WE/, Deep Power Down (DPD), ERASE, PROG, TMEN (test-mode enable), and CLK (for synchronous interface). Data-in DIN <0:N> typically comprises 16/32/64/128/256 data-input pins. Data-out DOUT <0:N> typically comprises 16/32/64/128/256 data-output pins. Address pins ADDR <N:0> are for addressing the memory array. Power supplies typically include V_{ddcore} (such as 1.x V) and V_{ddIO} (such as 1.8 or 2.5 V). Test pins include a test-enabling pin and test analog-bias pins.

The chip-operation modes include configuration bit load, read, sector or chip erase, word or page program, NVR-sector enable (NVR sectors are used for storing fuses, user configuration bits, manufacturing info), as well as testing functions. The configuration-bit load is used to configure various modes and to provide trimming fuses for the flash macro. Redundancy function, mainly sector redundancy, is used to improve yield. The error-correction code (ECC) function is used to improve the reliability of the macro, e.g., correcting tail bits during lifetime operation.

Erase operation, as shown in Fig. 5.23, uses the target erase voltage and erase timing for the selected sector as per Tables 5.2 and 5.3 for ESF1 and ESF3, respectively. Program operation, as shown in Fig. 5.24, uses the target program voltages, program current and program timing for the selected memory cells as per Tables 5.2 and 5.3 for ESF1 and ESF3, respectively.

To manage power, various chip-power modes are designed such as stand-by, deep power down, various read modes such as high-speed or low-power read, and variable word programming. Auto low-power mode is used to shut down unnecessary circuit blocks after a sensing cycle is completed. Transition from auto low-power into active read mode does not cost any additional access timing. Design for low-power employs techniques of block de-selection, multi-V_t, channel-L modulation, bulk modulation, source bias, and local and/or global power switch.

Fig. 5.23 Erase-operation path

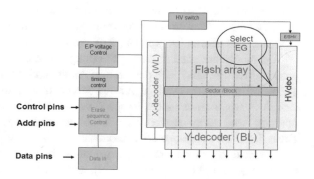

Fig. 5.24 Program-operation
path

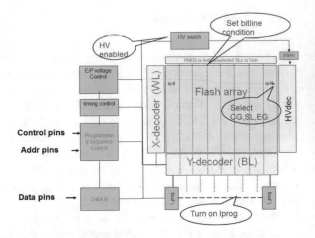

Table 5.1 Chip-operation modes

Chip-operation modes	Description
Configuration bit load	Loads configuration bits into the macro
Sector erase	Erases the selected array sector
Chip erase	Erases the whole array
Word program	Programs the selected word
Read	Reads the selected data word
Low-power read	Reads in low-power mode
E/P verify	Verifies data after an E/P pulse
Pre-program	A pre-program pulse to pre-condition the cells
Stand-by	Stand-by mode, i.e., ready for P/E/RD
Deep power down	Deep power mode, i.e., lowest power
Test modes	Access macro for testing

Erase and program (E/P) verify is used to enhance the reliability (endurance) and optimal E/P timing by reducing total stress and timing condition during E/P. Pre-program can also be used to improve endurance by softly programming the memory cells before actual programming, i.e., reducing electric field stress during program. The details for pre-programs are provided in Sect. 5.4.

Test modes are used to ensure the integrity of the flash product throughout the product lifetime. They includes various stress modes for the memory array and the peripheral circuits as well as various screen modes for marginal bits. They also includes various mass modes to speed up production testing and various modes to optimize the macro performance such as bias and timing on-chip trimming. Table 5.1 shows chip-operation modes and their descriptions.

A typical low-power macro performance for a 16-Mb macro with 32 IO width dual-voltage supplies (1.2 and 2.5 V), and 8-kbit sector size is as follows: read performance 50-MHz speed, I_{cc} read 3 ma, I_{cc} program 2.5 ma, I_{cc} erase 2 ma, I_{cc}

power down <0.5 μA 25 °C, sector-erase time 10 ms, chip-erase time 10 ms, word-program time 10 μs, data retention >10 years, and endurance 100 K cycles.

A typical high-speed macro performance for a 16-Mb macro with 144 IO width dual-voltage supplies (1.2 and 2.5 V) and 16-kb sector size is as follows: read performance 100-MHz speed, I_{cc} read 30 ma/×144, I_{cc} program 2.5 ma/×32, I_{cc} erase 2 ma, I_{cc} power down with power gating <2 μA 25 °C, sector-erase time 10 ms, chip-erase time 10 ms, word-program time 10 μs, data retention >10 years, and endurance 10/100 k cycles.

5.3.3 Array Architecture

A typical operation table of the ESF1 and ESF2 memory-array sector is shown in Table 5.2 and of the ESF3 memory array in Table 5.3. An example schematic of array operation is shown in Fig. 5.25 for ESF1 and of ESF3 in Fig. 5.26. The unselected source line is floating at a bias voltage during programming to reduce bit-line leakage. For ESF3 technology, all control gates are biased in read mode for enhancing speed from an increase in cell current. For some ESF3 applications erase gates are bias in read to enhance the cell current.

Table 5.2 Array-sector operation for ESF1 and ESF2

	BL		WL		SL	
	Selected	Unselected	Selected	Unselected	Selected	Unselected
Erase	0	0	V_{ERASE}	0	0	0
Program	I_{PROG}	V_{INH}	V_{WLPROG}	0	V_{SLPROG}	FLT
Read	V_{BLRD}	0	V_{WLRD}	0	0	0
Stand-by	0	0	0	0	0	0

Table 5.3 Array-sector operation for ESF3

	BL		WL		CG-selection sector	
	Selected	Unselected	Selected	Unselected	Selected	Unselected
Erase	0	0	0	0	0	0
Program	I_{PROG}	V_{INH}	V_{WLPROG}	0	V_{CGPROG}	V_{CGRD}
Read	V_{BLRD}	0	V_{WLRD}	0	V_{CGRD}	V_{CGRD}
Stand-by	0	0	0	0	V_{CGRD}	V_{CGRD}

	CG-unselectors	EG		SL	
	Unselected	Selected	Unselected	Selected	Unselected
Erase	V_{CGRD}	V_{ERASE}	0	0	0
Program	V_{CGRD}	V_{EGPROG}	0	V_{SLPROG}	FLT/bias
Read	V_{CGRD}	$V_{EGRD}/0$	$V_{EGRD}/0$	0	0
Stand-by	V_{CGRD}	$V_{EGRD}/0$	$V_{EGRD}/0$	0	0

Fig. 5.25 ESF1/ESF2
Array-operation schematics

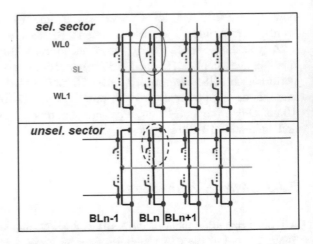

Fig. 5.26 ESF3
array-operation schematics

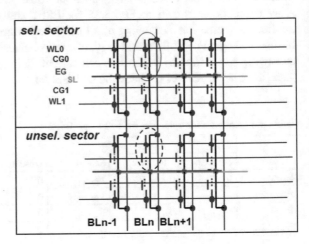

Typical values for ESF1 array operation are as follows: $V_{ERASE} = 11$–13 V, $I_{PROG} = 2$–2.5 μA, $V_{INH} = 2.0$–2.5 V, $V_{WLPROG} = 1.2$–1.5 V, $V_{SLPROG} = 9$–10 V, $V_{BLRD} = 0.8$ V, and $V_{WLRD} = 2.5$ V.

Typical values for ESF3 array operation are as follows: $V_{ERASE} = 11$ V to 12 V, $I_{PROG} = 1.0$ μA, $V_{INH} = 1.5$–2.5 V, $V_{WLPROG} = 0.8$–1.5 V, $V_{CGPROG} = 10$–11 V, $V_{EGPROG} = 4$–5 V, $V_{SLPROG} = 4$–5 V, $V_{BLRD} = 0.8$–1.2 V, $V_{WLRD} = 1.2$–2.5 V, $V_{CGRD} = 2.5$ V, and $V_{EGRD} = 1.2$ V.

Figure 5.27 shows a single-plane array organization suitable for a low-density or low-cost small-area macro. It consists of N array sectors (SEC0-SECN), redundancy sectors (RED), info row sectors (IFO), and a global-reference sector (GREF). An array sector SEC consists of M memory bytes that are erased at the same time, for example, 256 bytes for a smart-card application and 8 kB for a consumer application. Figure 5.28 shows an exemplary array sector with 4-kb sector size consisting of 8 rows with each row consisting of 64 bytes. Figure 5.29 shows an

Fig. 5.27 One-plane array organization

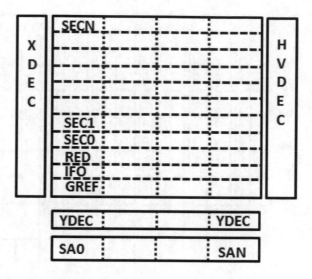

Fig. 5.28 Array-sector organization

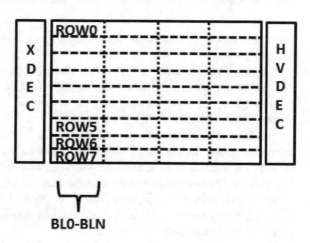

exemplary array sector with 2-kb sector size consisting of 4 rows with each row consisting of 64 bytes.

The array-sector organization, as shown in Figs. 5.28 and 5.29, is such that to satisfy disturb requirements resulting from a high-voltage condition applied to unselected memory cells in a selected sector during non-volatile operation as shown in the operating table. It includes same-row disturb (punch-through row PTR, disturb time = (#bytes per row) * t_{prog}, e.g., = 64 * 10 μs = 640 μs), column disturb (punch-through column PTC, disturb time = (#rows per sector-1) * t_{prog}, e.g., = 7 * 10 μs = 70 μs), and reverse-tunneling disturb (RT, disturb time = (# rows-1) * #bytes per row * t_{prog}, e.g., = 7 * 64 * 10 μs = 4.48 ms) from different rows in a same-selected sector. The program disturb is typically worse at higher

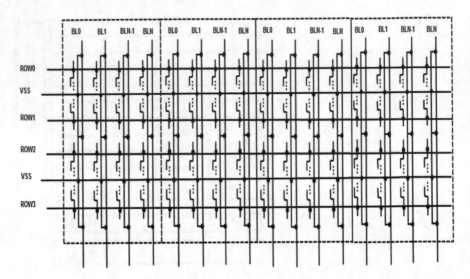

Fig. 5.29 Schematics of four row-array sector organization ESF1

temperatures; hence, disturb die screening is extensively done at high temperature at wafer sort.

The global reference sector (GREF), physically located next to the main array, is used to provide a reference level for sensing based on the reference memory cell. Dummy rows and columns surround the reference cells to ensure reference cell physical formation and electrical uniformity.

The info row sectors are provided to manufacturers to store manufacturing information, such as tracking info, and to users to store user information such as system fuse bits. The info sectors include configuration bits to configure the macro, fuse information to maximize the macro performance by trimming various bias and timing levels, redundancy information to enhance yield, security information for secured memory access such as encryption code, memory block access tags, macro password, and other information and data.

The redundancy sectors are used to replace defects in random bits, single rows, or sectors. They are activated at test, power-up, or system command. Column redundancy is at times used to replace column defect.

An example of array architecture suitable for a large sector size is shown in Fig. 5.30 with a segmented high-voltage array-plane organization. This array plane typically consists of 512 rows by 4/8 k columns. This architecture is generally more applicable for consumer and industrial specifications. The word line is much longer than that in Fig. 5.3a to accommodate a larger sector size, and the high-voltage decoder is shared across the shown two array planes. Hence, the row decoder and high voltage decoder area is much reduced at the expense of a longer word-line delay. Sector size is, for example, 16 kb with four rows by 4 k columns per sector.

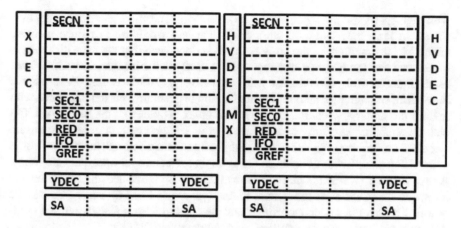

Fig. 5.30 Segmented HV-array organization: longer word line (medium speed), large sector size, small area

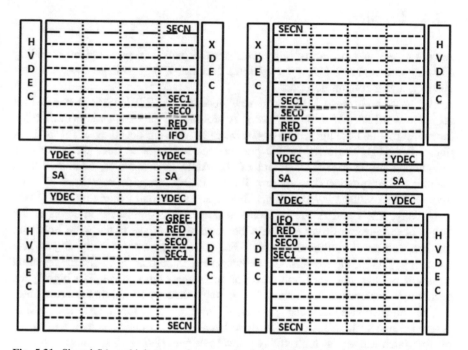

Fig. 5.31 Shared SA multiple-array plane organization: high speed, optimal area

An example of array architecture suitable for low-power, high-speed performance is shown in Fig. 5.31 with shared sensing architecture with smaller array-plane segmentation. This array plane typically consists of 256 rows by 4 k columns. This architecture is generally more applicable for automotive specification with high-speed and high-temperature requirements. The array plane is smaller than shown in Figs. 5.27 and 5.30 (shorter bit line and wordline) to reduce parasitic effect. Metal strapping for the array terminal is done such that the parasitic effect is minimized and the EM rule observed. M1 is used for the bit line; M2 and M3 are used for the control gate and erase gate, respectively; and M4 is used for the source line. Metal strapping for automotive application is generally relaxed compared with that for consumer application. Due to the complex geometry of the SL/EG/WL/CG pick-up strapping structure, careful layout with OPC correction is used to ensure electrical uniformity of the adjacent cells. Design for test is implemented to ensure optimal screening-stress effect from the strapping structure for operating-life usage. A read decoder (XDEC) is placed in the middle to optimize decoding interconnect for speed. The sensing is differential to maximize speed and dynamic to minimize power. This requires accurate timing and bias control for equalization and comparison period over process, voltage, and temperature (PVT).

5.3.4 Decoders

Access to the memory array is by word access, which is provided by decoding circuitry that includes a row (word line) decoder, a high-voltage row or sector decoder, and a column decoder.

Figure 5.32 shows a high-speed row decoder for ESF1/ESF2 that provides access to word lines of a memory array. A pre-decoder block is used to decode the incoming-row address lines, e.g., A <0:N>, to provide pre-decoding of row address lines XPA-XPD. These XPA-XPD pre-decoded row address lines are inputs to a low-voltage AND gate. The output of the AND gate together, combined with another set of pre-decoded address lines XPZ0-N, provide row lines WL0-N through a low-voltage inverter driver. For first- and second generation SuperFlash technology, ESF1 and ESF2, an isolation transistor is connected in series to the output of the inverter driver to isolate the erase high voltage in the word line from the low-voltage decoding transistors. For the latest generation of SuperFlash technology, ESF3, the row decoder shown in Fig. 5.33 does not need the isolation transistor because the word line does not encounter high voltage. The high-speed decoding is achieved through using low-voltage circuitry for driving the row lines.

The high-voltage decoder is shown in Fig. 5.34 for ESF1 and ESF2. A low-voltage latch LVLATCH is used to latch the sector-enable signal. The output of the latch drives a high-voltage level shifter HVLS. The output of the high-voltage level shifter is used to enable or disable the word-line WL, the source line SL, or the coupling gate CG in erase, program, or read. The erase word-line decoder WLERDEC includes a current limiter biased by EGHV_BIAS to control current supplied to the word line during erase. The source-line SLDEC includes a Kelvin decoder for accurate SL-voltage delivery. The source-line SLDEC also provides a bias for un-selected source lines during programming to limit column-leakage current.

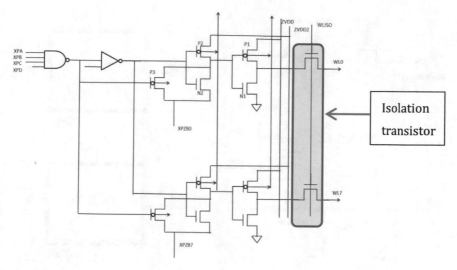

Fig. 5.32 Row decoder for ESF1/ESF2

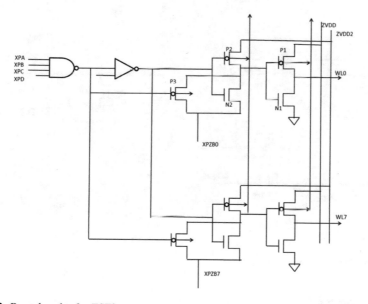

Fig. 5.33 Row decoder for ESF3

The high-voltage decoder is shown in Fig. 5.35 for ESF3. EG decoder EGDEC is similar to WLERDEC of ESF1/ESF2. SLDEC is similar to that of the ESF1/ESF2. Coupling gate decoder CGDEC provides voltages to the CG gate during programming, erase, read, and stand-by. During read, CGDEC and/or EGDEC provide a bias voltage on the coupling and/or erase gates to increase the cell current.

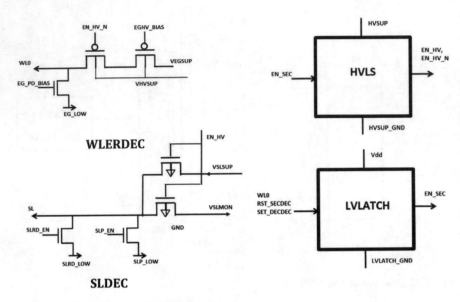

Fig. 5.34   HV decoder for ESF1/ESF2

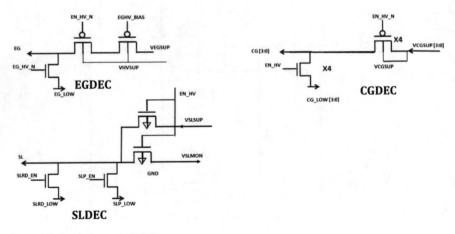

Fig. 5.35   HV decoder for ESF3

5.3.5   Sensing

Sensing is one of the most critical operations for a flash-memory system. It demands rigorous and methodical design techniques to ensure the utmost product reliability and robust functional safety, especially in severe environments such as in automotive applications. For SuperFlash memory, the sensing starts with a design-margin analysis that consists of cell-margin analysis, array analysis, and sensing circuit-variation analysis. Figure 5.36 shows a cell-window analysis.

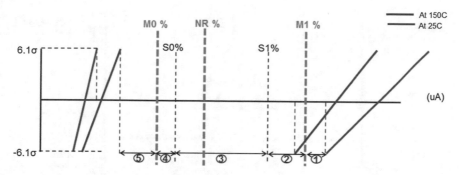

Fig. 5.36 Cell-margin window

The sensing cell window considers six sigma distribution of I_{r1} (erased cell current [typically ~20–35 µA]) and I_{r0} (programmed cell current [typically <1 µA]) array leakage at high temperature (typically <0.5 µA), endurance degradation (no. 1 [typically <1 µA/10 K cycles]), data retention (no. 2, no. 5 [typically <1 µA]), and trap relaxation (no. 4 [typically <1 µA]). The final cell window (no. 3) takes into account the sense-circuit mismatch, reference mismatch, array parasitic, and required at-speed delta input voltage for the comparator. The array analysis takes in account the array-column leakage at high temperature (e.g., 170 °C), bit-line RC parasitic, word-line RC delay, source-line RC delay, and source-line and bit-line IR-voltage decrease. The sensing-circuit variation must examine the mismatch of the reference circuit path, the IO circuit path, and the offset of the sensing comparator over the PVT (process, voltage, and temperature). The sensing-circuit variation also includes a statistical model analysis.

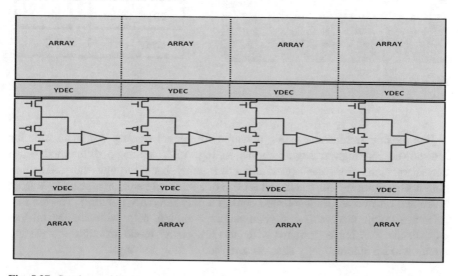

Fig. 5.37 Sensing architecture

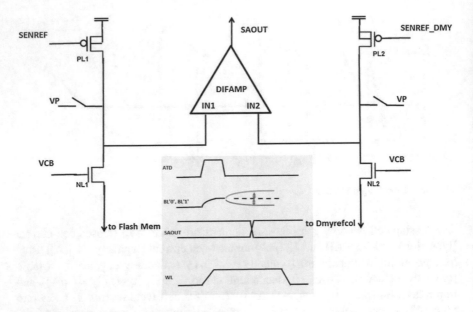

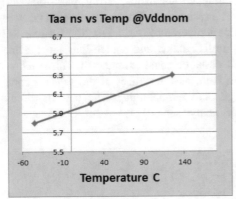

Fig. 5.38 Sensing amplifier

@Vddcore, Temp	SA offset sigma (mV)
1.21V_125C	4.8
1.21V_-40C	4.5
0.76V_125C	10.6
0.76V_-40C	5.8

The sensing architecture uses a dynamic differential-sensing system suitable for low power and high speed as shown in Fig. 5.37. Inputs to differential sense amplifiers come from the upper and lower array bank through the y-decoding circuit. One input is from a selected bit line of a selected bank, and the other is from a selected bit line of an un-selected bank (dummy reference column). The dummy reference column serves as a symmetrical coupling path with similar bit-line capacitance. Sensing is designed with auto low-power mode, meaning the sensing circuit will be automatically shut off after completing sensing.

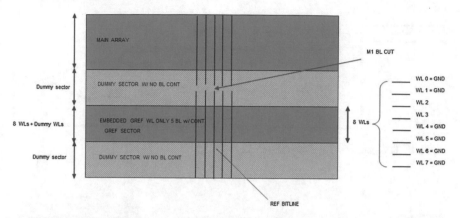

Fig. 5.39   Reference-array sector

The low-power and high-speed wide-swing dynamic-sense amplifier is shown in detail in Fig. 5.38. The insert shows the sensing-timing sequence. An ATD pulse pre-charges a selected bit line and equalizes the inputs to the comparator. The comparator will sense the developed delta voltage shortly after to output a "1" or "0" depending on the erased or programmed state. A reference current I_{ref} is provided from the PMOS transistor PL1, which is mirrored from a global reference circuit coupled to a reference memory cell. The NMOS transistor NL1 serves to couple to a selected memory cell-current I_r through a selected column. Thus, the input to the comparator IN1 is the sensing node, which is proportional to the delta current between the I_{ref} and the I_r. This sensed node is a high-impedance node; hence, the signal swing is wide, and it provides a good voltage margin for the dynamic comparator. A dynamic instead of DC comparator is used to save power. Monte Carlo offset analysis for the sense amplifier is shown at approximately 10 mV for a worst-case scenario. An 8-Mb high-speed macro access time at V_{ddnom} is <6.5 ns at 125 °C.

The dummy reference side consists of transistor PL2 and NL2 serves as a voltage-holding node for other input IN2 for the comparator.

A reference memory sector is used to provide a reference memory cell as shown in Fig. 5.39. The reference sector is exactly the same as the main memory array with interconnect isolated to not interfere with the operation of the main array. The reference memory cell is located in the middle of the reference sector with surrounding rows and column serves as a dummy to ensure reference-cell uniformity.

5.3.6   High-Voltage Generation and Regulation

The high-voltage generation and regulation block has two major blocks: The EG/CG HV BLK and the SL HV BLK. The EG/CG HV BLK provides the EG

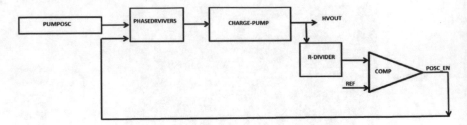

Fig. 5.40 HV generation and regulation

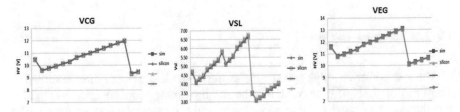

Fig. 5.41 HV-data simulation versus silicon at nominal condition

erase voltage and the CG program voltage. The SL HV BLK provides SL source-line programming voltage.

HV generation and regulation, as shown in Fig. 5.40, uses an on/off pump-clock driver scheme for low-power purpose. A low-power divider network steps down the high voltage to a nominal voltage, and this stepped-down voltage is compared with a target-trimmed reference voltage. Once the pumped high voltage reaches a target, the output of the comparator disables the clock driver. An RC filter network acts to reduce the output ripple voltage. The clock control is synchronized with the clock phase to reduce the voltage ripple.

SL regulation further includes a compensation technique to maintain the accurate voltage across varied programming currents over PVT. The compensation technique employs a compensation loop to keep the SL voltage constant at memory array.

The charge pump uses a Dickson-type charge pump circuit with two-phase clock drive, V_t-cancellation technique, and native high-voltage MOS pass transistor. The charge-pump capacitor is an area-efficient composite capacitor combining a MOS and a specialized MOM capacitor to sustain necessary high-voltage during E/P operation.

Figure 5.41 shows excellent matching for HV-trim range-data simulation versus silicon at nominal condition V_{dd}-nom and 25 °C.

5.3.7   Design for Test and Reliability

Design for test is implemented to ensure product quality at a nethermost test cost through defect and disturb screen modes, endurance screen, operation margining, mass modes, bias trimming, E/P high-voltage trimming, and read-time trimming.

Design for test is implemented to ensure that product specification is met with margin and so that memory-array access is available for in-field failure analysis. Voltage monitors, charge-pump output monitor and forcing, external V_{PP} (HV) supply for mass modes and stress modes, cell-current measurement, program-current measurement, and at-speed read test are implemented.

Automotive designs must have a zero-defect goal and functional safety target. Because of this, the ability to control and observe internal critical timing—such as read timing and erase and programming biases—are meticulously implemented with minimum area overhead and no performance impact. Extensive test modes are used to screen all known latent defects. Power sequencing for dual supplies are carefully observed to ensure voltage integrity, especially cross-domain voltages during power up and power down for internal circuits.

Design for reliability is done to ensure reliable operation during 10 years with a specified endurance over a range of temperatures. A high-voltage stress-reduction design technique is used to ensure high-voltage transistor reliability especially at high temperature such as at T_j 170 °C. Certain operational modes are used to enhance the endurance of elements such as pre-program and erase and program retry. Pre-program programs the memory cells with a shorter program pulse ($\times$ two to four shorter times) and lower (e.g., by 3–4 V), thus programming voltages to pre-condition the memory cells to reduce the stress for the subsequent program operation. Erase and program retry executes shorter pulses with incremental voltage pulses for shorter total E/P time and optimal field-stress conditions during E/P operation. Redundancy is used to screen out tailed bits identified by extensive screening during wafer sort. ECC is used to further enhance product reliability, especially for automotive applications.

5.4   SuperFlash Memory-Reliability Fundamentals

5.4.1   Reliability at High Temperature

The oil-exploration industry has successfully used high-temperature electronics, with SuperFlash technology, on drill-head monitors with operating temperatures ≤ 220 °C while also under severe conditions of mechanical shock and vibration. Although automotive high-temperature electronics are designed to operate under somewhat less harsh conditions, they must meet cost-effective and higher volume production demands, zero-ppm failure tolerance, and a significantly longer device lifetime. In addition, there is an increasing demand for high-performance, low-power electronics.

Modern-day cars use microcontrollers (MCU) in most of automotive systems including engine controls and powertrain control electronics. The junction temperature of automotive MCUs can range from –40 to 175 °C.

Table 5.4 shows the most popular embedded flash technologies being used by leading MCU chipmakers. Split-gate embedded flash-memory solutions are becoming one of the main choices of designers for high-performance as well as low-power MCU applications. Split-gate memory cell can offer wide operating window between programmed and erased states thanks to its immunity to over-erase issues, highly efficient hot electron injection programming mechanism [24], and simpler design for higher performance and long-term reliability.

Because any charge-storage memory cell is scaled down, the single-electron stochastic (statistical) behaviors can have a severe impact on memory performance, reliability, and technology scaling. Generally, as floating-gate non-volatile memory technology has continued to be scaled down, it has suffered from a low electron-count issue. However, the ESF3 cell is almost unaffected by low electron-count effects down to 28-nm node as shown by the ESF scaling trend in Fig. 5.42 a. The chart illustrates the SuperFlash memory-scaling trend in terms of stored electrons between programmed and erased states. At 28-nm technology node, the programmed and erased states in the ESF3 cell are separated by almost

Table 5.4 Popular embedded flash solutions used in modern MCU products

	1.5T SuperFlash	1.5T TFS	1.5T Poly FG	1T Poly FG	1.5T ONO
Scaling and R&D	40–28 nm	90–40 nm	40 nm	40 nm	40–28 nm
Key drivers	SST and licenses and 100+ users	FSL	Infineon	ST Micro/FSL	Renesas
V_{CC}=1.2 V fast read, LP	Yes	–	Need $V_{read} > 2$ V	Need $V_{read} > 2$ V	Yes
P/E	SSI-HEI/FN to EG 100k–1M	SSI-HEI/FN to CG 10–100K	SSI-HEI/FN to channel, 10–100k	CHEI/FN to channel 100k	SSI-HHI/FN to SiN, 10–100k
1–0 V_T window	>10 V	~3 V	~6 V	~3–4 V	~4 V
Scaling challenges	FGOX, HV/CG EG	Low Q, O-NC-O, HV	SGOX, FGIOX, HV/ V_{read}	Low Q,FG CD (PD), FGOX HV/V_{read}	Low Q, ONO

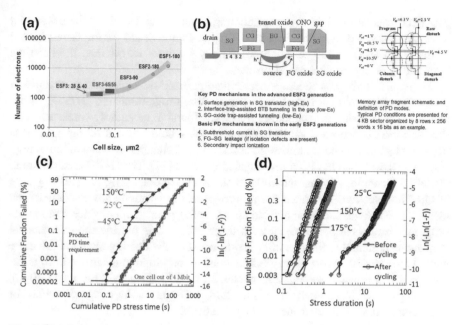

Fig. 5.42 a SuperFlash memory-cell scaling. **b** Schematic cross-section of basic structural unit of the third-generation SuperFlash including two cells, memory-array schematic, and definition of PD modes. **c** Program-disturb time-to-failure distributions for 4-Mb array. **d** ESF3 PD distributions for a 4-KB sector

1000 electrons. The coupling-gate (C_G) threshold-voltage (V_T) delta between programmed and erased states exceeds 10 V [25].

In addition to the large operating window between the programmed and erased states, split-gate ESF3 memory cell can also offer scalable supply voltage for read operation down to logic power supply (≤ 1.1 V), thus making it a well-suited solution for automotive applications that demand high-performance, high-reliability, and low-power solutions.

Figure 5.42b summarizes key program-disturb (PD) mechanisms in ESF3 arrays. Typical PD conditions are presented for a 4-KB sector organized by 8 rows × 4096 columns as an example. ESF3 array architecture avoids or effectively suppresses PD initiated by sub-threshold current through the select-gate (SG) transistor (mechanism 4), leakage through FG–SG isolation (mechanism 5), and impact ionization at the source area (mechanism 6). The immunity of ESF3 technology to these basic PD mechanisms have been highlighted and described previously [26, 27].

Figure 5.42c represents program-disturb distributions for a 4-Mb array based on ESF3 technology at 40-nm technology node. ESF3 cell on 40-nm technology node provides good program-disturb immunity across the wide junction temperature range of −45 to 175 °C. Programming and program-disturb conditions in the chart are similar to the ones shown in Fig. 5.42b. For these measurements, the coupling gate (CG) was decoded during programming to minimize disturb on the unselected rows. As evidenced from the chart here, time-to-failure (TTF) for the fastest bit out of the 4-Mb array at 150 °C is approximately 40 times longer than the PD time required by an automotive MCU product.

Program disturb (PD) becomes accelerated at increased temperatures because of greater sub-threshold current. A high-temperature program-disturb (HTPD) mechanism in ESF3 at 40-nm technology node was found to be entirely consistent with that identified in the previous technology nodes such as 90 and 65-nm [28]. It is triggered by electron–hole surface generation in SG channel with activation energy (Ea) of ~ 0.5 eV and has been investigated in detail [27]. The HTPD mechanism in the ESF3 technology is insusceptible to stress due to erase/program (E/P) cycling because hot electrons generated during programming operation are positioned well inside the FG channel, that is, outside the area of electron supply, which can initiate program disturb. As for erase-induced degradation, it is outside the cell-channel area and is confined to the tunneling oxide between the floating gate and the erase gate. Figure 5.42d shows a negligible change in program-disturb distributions measured before and after 5E5 E/P cycles on a ESF3 memory array using 40-nm technology node. As evident from the chart, ESF3 technology is well suited for high-temperature embedded memory applications.

5.4.2 SuperFlash Memory High-Endurance Characteristics

Smart-card applications require fast updates to the embedded flash memory. This requirement translates to fast erase and program times within just a few milliseconds for a typical memory sector size of 128 B–1 kB. In addition, many smart-card applications have high-endurance requirements ≥ 100 K cycles.

When a positive voltage (VEG) is applied to the erase gate (EG), the tunneling injection of electrons from the floating gate (FG) to the erase gate takes place at the FG corner thanks to an electric field-enhancement mechanism. Non-uniform electric field distribution in tunnel oxide during erase operation facilitates efficient forward electron tunneling from the FG corner and prevents reverse anode hole injection from the EG back to the tunnel oxide. Therefore, relatively thick tunnel oxide can be used in ESF3 technology, thus making it highly resistant to stress-induced leakage current (SILC) and therefore a reliable and robust memory solution [29]. Figure 5.43a illustrates the difference between tunneling-oxide band diagrams for ESF3 cell with FG corner-enhanced tunneling injection and stacked-gate NOR cell with FG-planar tunneling injection during erase operation. E_e is the electron energy that is transferred to a valence-band electron resulting in the generation of a hot hole at the anode surface with a possibility of its injection into the tunneling oxide, thus resulting in oxide-trap generation and eventually SILC formation.

Regarding the high-endurance requirement of smart cards, we measured the erase and program time for a series of typical multi-Mb memory arrays across

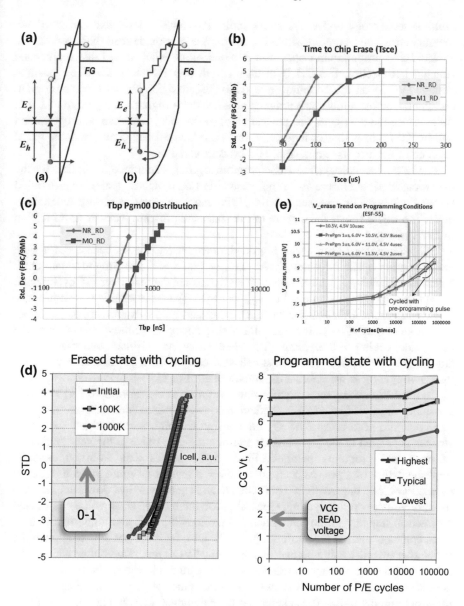

Fig. 5.43 a Simplified band diagrams during erase operation for *a* planar geometry (stacked-gate NOR cell) and *b* corner (tip) geometry (ESF cell). **b** Typical ESF3 memory-array erase time. **c** Typical ESF3 memory-array program time. **d** A 32-Kb sector-operation window from the erase (cell current) and program (CG VT) sides. **e** Effect of different program-operation conditions on an increase in erase voltage

multiple technology nodes. As an example, Fig. 5.43b shows that the multi-Mb memory array can be erased within 0.1 ms to pass user-mode read (NR) conditions.

The ESF3-programming operation is based on an efficient and fast source-side channel hot-electron (SS-CHE) mechanism that is further assisted by positive voltage (VCG) on the coupling gate (CG). Figure 5.43c shows time-to-program distribution across a multi-Mb array using a low cell-programming current of 1 μA. As evidenced in the figure, the slowest bit of the memory array can be programmed within 500 nS to pass user-mode read, thus producing a sufficiently reliable guard band and tolerance to temperature and process variations.

For ESF3 technology, the erase/program cycling endurance is limited by the operation window closure from the erase side (logic "ONE" state). Figure 5.43d shows typical cycling results for the ESF3 memory array at 65-nm technology node. The figure shows the closure of the erase and program window as a function of E/P cycling.

ESF3 technology excels with high-endurance characteristics because of a wide erase-program operation window. As seen in Fig. 5.43d, the program state characterized by CG V_{Tmin} exceeds +5 V, which does not degrade over E/P cycling and is kept well above the read level of $V_{\text{CG}} = 1.8$ V. The erase state is shown in terms of read-current distribution for the fresh array and gradually moving to the lower current level after E/P cycling, thus illustrating operation-window closure from the erase side during E/P cycling. The data show an intrinsic sector-endurance capability >1 M cycles. The lowest cell-read current for a 32-kb sector after 1 M cycles of cycling is well above the reference read level.

To meet high-endurance requirements toward 1 M erase/program (E/P) cycling capability on a product level, a meticulous design optimization of erase and program operation can be applied. Optimal design techniques can mitigate peak field stresses generated in the tunnel oxide and floating-gate oxide at the very beginning of erase and program operation. Figure 5.43e shows data from an example implementation of a pre-program pulse before a regular program pulse in order to reduce FG-oxide degradation. This scheme consequently slows down the increase of erase voltage during cycling. Generally, the increase of erase voltage is caused by electron trapping in the tunnel and FG oxides [25]. It is also important to note that implementation of the short pre-program pulse makes it possible to shorten the main program pulse using higher than the typical VCG program voltage without the risk of introducing damage to the FG oxide. Figure 5.43e shows the results from a high-speed program operation with a total time of 3 μs including a 1-μs pre-programming pulse. The scheme has been demonstrated on a design with ESF3 cell on 55-nm technology node, which exceeded 500-k cycling.

Typical program conditions are based on a single-pulse operation: $V_{\text{CG}} = 10.5$ V, $V_{\text{EG}} = V_{\text{SL}} = 4.5$ V, $V_{\text{SG}} \sim 1$ V, program current = 1 μA and 10-μS pulse duration: The corresponding data of V_{erase} cycling trend are shown in blue. The other data correspond to cycling with pre-programming pulse showing reduced V_{erase} degradation and improved endurance.

With efficient and fast erase and program operation, energy consumption by ESF3 flash designs can be greatly reduced to meet growing demands for extended battery-lifetime applications. Another important feature of the ESF3 memory cell is its scalable and compatible select-gate (SG) transistor with low-voltage CMOS logic transistors. There is no need for high-voltage pump circuitry to read the ESF3 cell.

ESF3 technology with low-voltage read operation, fast erase and program operation, and high E/P cycling endurance capability well above 100 k is an excellent fit for smart-card and IoT applications.

5.5 SuperFlash Memory-Technology Scaling

Recent advances in automotive-electronics technology and the explosive growth of semiconductor devices being used in the Internet of Things are expanding the applications of embedded non-volatile memory-based microcontrollers. Such demand continues to push the scaling of industry-proven embedded NVM technologies and the search for new types of embedded NVM technologies that can easily be integrated into various CMOS-baseline processes.

As publicly known, not all existing NVM technologies currently used in the embedded-application space can be scaled cost-effectively and reliably to the advanced-process nodes such as 40 and 28 nm. Table 5.5 summarizes the scalability of various eNVM technologies [30–39]. As shown, split-gate (SG) flash-memory solutions dominate the scaling path in the eNVM landscape. The most versatile and widely used eNVM technology in the advanced embedded-flash market currently is SST's SuperFlash technology, specifically ESF3, with 40 nm available for production and 28 nm [31] in the development stage. The logic-compatible integration of ESF3 technology into the baseline processes has enabled its adoption at various leading manufacturing partners and allowed their existing logic libraries to remain unchanged. The worldwide foundry adoption of this technology also assures customers of its availability and provides competitive cost benefits. Thus, this section focuses on the current status of SG flash technologies with the performance and reliability of ESF3 described in detail.

Despite the aggressive scaling of cell size, the performance and reliability of the memory array are not compromised. This technology therefore continues to enable a wide range of modern embedded non-volatile applications, e.g., smart-card devices, that require very high endurance and low power; automotive microcontrollers that require zero data retention error with fast access time; and IoT devices that require extremely low power consumption.

In this section, performance and reliability data from flash arrays of multiple megabits fabricated in 90-, 55-, and 40-nm logic processes are summarized. With the same process and an identical bit cell, flash macros for different application purposes can be designed for each node. Regardless of the differences in speed requirements, the silicon results have many features in common: excellent operation window for the wide temperature range of −45 to 150 °C coupled with high endurance and robust data-retention capability. Silicon data from an ESF3 cell

Table 5.5 Existing landscape of eFlash technologies

	1.5T ESF3	1.5T MONOS	1.5T HS3P	1.5T TFS	1T FG	2T FG	SONOS	2T pFlash
Key Drivers	SST and 100 +users	Renesas	Infineon	FSL/NXP	STM, NXP, …	NXP	CY, eMemor	ISSI, SMIC
40 nm	Yes	Yes	Yes	No	–	Yes	–	No
28 nm	Yes	Yes	–	–	–	–	–	–
V_{cc} read	Yes	Yes	–	–	–	–	–	–
V_t window	>10 V	~4 V	~6 V	~3 V	<5 V	<5 V	~4 V	–
P/E	100–500k	10k for Code 1M for Data	100k	100k	100k	100k	1–100k	100k
Automotive	Yes	Yes	Yes	-	Yes	Yes	–	–

fabricated at 28-m process node are exhibiting performance and reliability inherited from previous generations.

5.5.1 ESF3 Scaling

Figure 5.44 shows the scaling of the ESF3 cell with normalized cell size as a function of process technology nodes. As shown, cell size is linearly scaled with technology nodes down to 28 nm. Early work on critical dimension design with 28-nm process indicates that the ESF3 22-nm cell can achieve reasonable scaling. In addition, an initial concept to integrate the ESF3 cell on fully depleted silicon-on-insulator (FDSOI) has been proposed for process nodes ≤ 28 nm [40].

Maintaining the linear scaling of ESF3 requires innovation while ensuring the minimal impact on baseline process devices. As shown in Fig. 5.15, most of the ESF3 flash modules are integrated before logic process, whereas those modules such as shallow-trench isolation (STI) shared between flash and logic devices, are

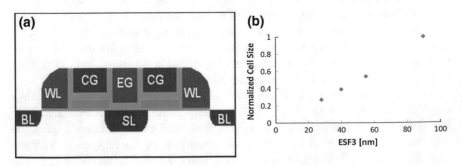

Fig. 5.44 ESF3 Scaling: **a** ESF3 cell structure and **b** normalized cell size as a function of technology node

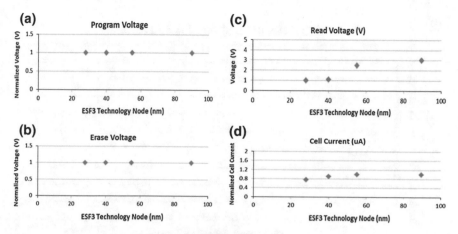

Fig. 5.45 ESF3-voltage scaling: **a** normalized CG-programming voltage, **b** normalized erase voltage, **c** WL read voltage, **d** normalized cell current

kept unchanged. With 90 and 110-nm nodes, only lateral scaling (cell width and length directions) needs to be addressed. Floating-gate (FG) poly is formed to self-align to active in the x-direction and to the coupling gate (CG) in the y-direction. It is important to make sure that the critical dimension of the memory cell does not violate the logic-design rules. For example, the spacing of FG–FG poly on STI is optimized for process tolerance and electrical cross-talk. In the cell-length direction (y), the FG and select-gate (WL) channels and the source active are scaled to optimize for read and programming performance. The rest of the ESF3 cell dimensions are scaled to logic rules.

Beginning with 65-nm process technology, logic-gate poly thickness scales to approximately 100 nm. This poses challenges not only to the integration of the ESF3 memory cell in the vertical direction but also to maintain the high-voltage (HV) transistor's performance. To overcome certain process integration hurdles, such as chemical mechanical polishing (CMP) steps, ESF3 cell height scales to retain its performance and reliability.

As discussed in [41], despite aggressive scaling, high voltages used in erase and program operations do not increase (Fig. 5.45a, b). As shown in Fig. 5.45c, d, read voltage reduced with advanced nodes and erased cell current reduces slightly. The reduction of read voltage improves power, and a minor design improvement would compensate for the slight reduction of cell current to maintain the read-access time.

Figure 5.46a, b show the distribution of normalized cell currents in both the erased and programmed states for 55 and 40-nm cell arrays, respectively. Despite the cell-size reduction from previous nodes, the ESF3 40-nm cell continues to show typical erased and programmed cell current >30 μA and near 0 μA, respectively, at room temperature. As part of the scaling, the dielectric under the WL transistor in ESF3 cell can either be the IO (2.5 or 3.3 V) or the core gate oxide. Using the core gate oxide for the WL transistor allows the array-read operation with core supply

(a)

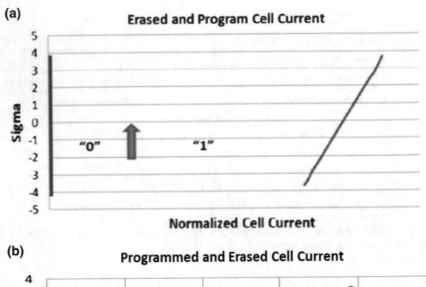

(b)

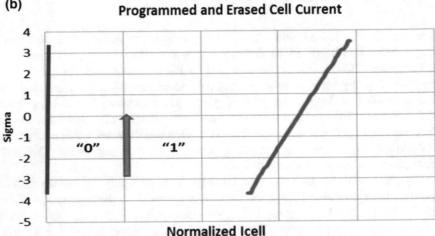

Fig. 5.46 Normalized ESF3 programmed and erased cell–current distributions for **a** 55-nm and **b** 40-nm array

Fig. 5.47 ESF3 55-nm array-byte programming time

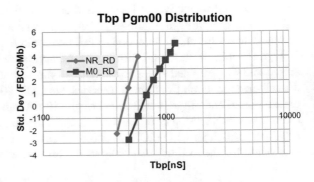

Fig. 5.48 Program-disturb immunity at 150 °C in ESF3 55 and 40-nm arrays

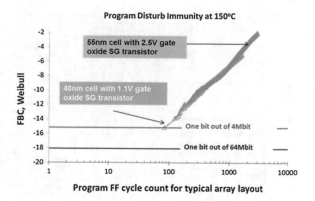

Fig. 5.49 ESF3 endurance: pre- and post-cycled erased cell–current distributions in **a** 55-nm and **b** 40-nm arrays

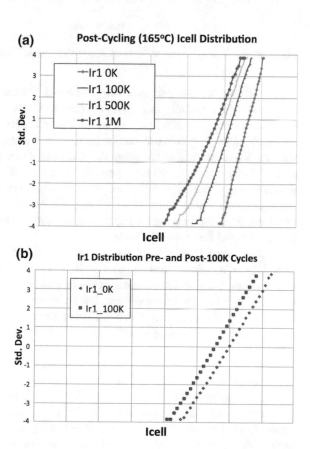

voltage. To overcome the smaller WL overdrive voltage V_{dd}, i.e., threshold voltage, the cell channel is optimized to maintain a good erased cell current distribution.

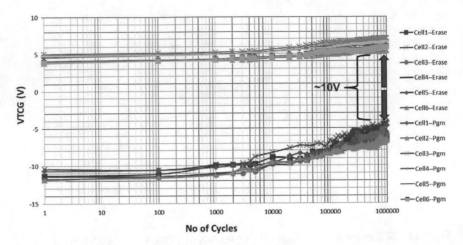

Fig. 5.50 ESF3 programmed and erased V_{tcg} window up to 1 M P/E cycles in 28-nm node

In addition to fast programming, where a typical cell can be programmed in >1 µAs, the immunity to program disturb in ESF3 cell is excellent. Figure 5.47 exhibits the program efficiency of the ESF3 cell arrays in 55 nm. As shown, the slowest byte in a typical array can be programmed in approximately 0.5 µs with the user mode. Importantly, the program efficiency does not degrade when the cell scales from 55 to 40-nm process. Figure 5.48 summarizes the ESF3 programming-disturb immunity for both 55 and 40-nm technologies. Despite the scaling of WL oxide from 55 to 40 nm, the program-disturb immunity remains excellent, thus enabling high-temperature operation in automotive applications with low power and fast read operation.

Because the EG is separated from the WL and HV gate, although sharing the same logic polysilicon, tunneling oxide between the FG and the EG is continually improved to provide the greatest possible endurance capability. Figure 5.49a, b reveal the erased cell-current distribution before and after P/E cycling. At the end of 100 K cycles, the typical erased-cell current decreases only approximately 2 µA in a 40-nm cell, thus leaving a wide margin between the tail bits and the reference level.

In 28-nm node, the ESF3 cell continues to scale while maintaining critical performance and reliability parameters. As shown in Fig. 5.50, a group of 28-nm cells reveals excellent endurance capability. After 1 M P/E cycles, the cells leave a wide margin of cell current between "program" and "erase" with a 10 V V_{tcg} window. With this performance trait, ESF3 will continue to serve a wide range of market needs from smart cards to high-end automotive micro-controllers that require high speed and high reliability.

5.6 Conclusion

SuperFlash memory has evolved during the last 27 years since its inception into an embedded flash-memory technology of choice for a wide variety of applications. Despite the challenges in integration, baseline process materials, and scaling, every new generation of SuperFlash memory performs better than the previous generation while retaining its excellence in terms of reliability. In addition to embedded applications for automotive, and industrial segments, split-gate SuperFlash memory technology has become a preferred choice for smart-card applications due to its fast program/erase time of just a few milliseconds, read access time, smaller sector architecture, low-power capability, and ability to achieve million-cycle endurance.

The ease of integrating the high-performance SuperFlash memory cell into baseline CMOS-logic process has made SST a strong and long-lasting partner with several of the world's leading foundries and IDMs, thus allowing SST's SuperFlash technology to be an attractive and dependable solution for designers of embedded flash systems.

Government regulations and consumers' desire for safety, fuel efficiency, and convenience will continue to drive the automotive semiconductor industry [42]. Also, the need for advanced near-field communication features will continue to stimulate the smart-card industry. With the help of its worldwide partners, SuperFlash memory technology will continue to find its home in billions of integrated circuits for many years to come.

References

1. http://voyager.jpl.nasa.gov/science/neptune.html
2. F. Masuoka et. al. A new Flash EEPROM cell using triple polysilicon technology, IEEE Technical Digest IEDM (1984)
3. Embedded Flash Memories for Nano-Scale VLSIs, chapter 6, Springer, Berlin
4. http://www.eetimes.com/document.asp?doc_id=1153458
5. http://ww1.microchip.com/downloads/pr_archive/en/en546947.pdf
6. B. Yeh, Single transistor non-volatile electrically alterable semiconductor memory device, U. S. Patent 5,029,130, 2 July 1991
7. S. Kianian, A. Levi, D. Lee, Y.-W. Hu, VLSI Symposium. Tech. Dig. **6A**, 71–72 (1994)
8. D. Lee, Self aligned method of forming a semiconductor memory array of floating gate memory cells and a memory array made thereby, U.S. Patent 6,329,685, 11 Dec 2001
9. V. Markov, X. Liu, A. Kotov, A. Levi, T. Dang, and Y. Tkachev., in *NVMTS* (2003), p. 23
10. L.Q. Luo, Y.T. Chow, X.S. Cai, F. Zhang, Z.Q. Teo, D.X. Wang, K.Y. Lim, B.B. Zhou, J.Q. Liu, A. Yeo, T.L. Chang, Y.J. Kong, C.W. Yap, S. Lup, R. Long, J.B. Tan, D. Shum, N. Do, J.H. Kim, P. Ghazavi, V. Tiwari, in *IEEE IMW* (2015), pp. 165–169
11. Y. Tkachev, X. Liu, A. Kotov, V. Markov, A. Levi, in *NVMTS* (2004), pp. 45–50
12. N. Do, in *IEEE ICICDT* (2016), pp. 1–4
13. M. Kamiya, Y. Kojima, Y. Kato, K. Tanaka, Y. Hayashi, in *IEEE IEDM* (1982), pp. 741–744
14. A. T. Wu, T. Y. Chan, P.K. Ko, C. Hu, in *IEEE IEDM* (1986), pp. 584–587
15. H.C. Sung, F.L. Tan, T.H. Hsu, Y.C. Kao, Y.T. Lin, C.S. Wang., in *IEEE Electron Device Letters*, vol. 26, no. 3 (2005), pp. 194–196
16. Y. Dong, W. Kong, N. Do, S. L. Wang, G. Lee, in *Solid-State Electronics*, vol. 54 (2010), pp. 579–581

17. Y. Dong, W. Kong, N. Do, S. L. Wang, G. Lee, J. Semiconductors **31** (2010)
18. X. Liu, V. Markov, A. Kotov, T. Dang, A. Levi, in *IEEE ICSSCIT* (2006)
19. A. Kotov, A. Levi, Y. Tkachev, and V. Markov, in IEEE NVMTS, 2002
20. Y. Tkachev, in *IEEE ICMTS* (2016), pp. 110–115
21. H. Om'mani, M. Tadayoni, N. Thota, I. Yue, N. Do, in *IEEE ICMTS* (2013), pp. 192–194
22. Y. Tkachev, X. Liu, A. Kotov, in *IEEE Transactions on Electron Devices*, vol. 59, no. 1 (2012), pp. 5–11
23. N. Do, in *IEEE IMW* (2016), pp. 8–11
24. J. Van Houdt, P. Heremans, L. Deferm, G. Groeseneken, and H. Maes, in *IEEE Transactions on Electron Devices*, vol. 39 (1992), pp. 1150–1156
25. A. Kotov, *Leading Edge Embedded Non Volatile Memories* (2015)
26. V. Markov, K. Korablev, A. Kotov, X. Liu, Y. B. Jia, T. N. Dang, A. Levi, in *IIRW Final Report* (2007), pp. 43–47
27. V. Markov, A. Kotov, *IEEE Trans. on Device and Materials Reliability*, vol. 14, no. 2 (2014) pp. 672–680
28. V. Markov, J. Kim, A. Kotov, in *Proceedings of the IEEE IMW* (2016), pp. 21–24
29. A. Kotov, in *MRS 2015 Fall Meeting, Symposium KK: Materials and Technology for Non-Volatile Memories* (2015)
30. Y.K Lee, B. Seo, T-K Yu, B. Lee, E. Kim, C. Jeon, W. Park, Y. Kim, D. Lee, H. Lee, S. Cho, in *IEEE IMW* (2014), pp. 75–78
31. L.Q. Luo, Z.Q. Teo, Y.J. Kong, F.X. Deng, J.Q. Liu, F. Zhang, X.S. Cai, K.Y. Lim, P. Khoo, S.M. Jung, S.Y. Siah, D. Shum, C.M. Wang, J.C. Xing, G.Y. Liu, L. Tee, S.M. Lemke, P. Ghazavi, X. Liu, N. Do, in *IEEE IMW* (2016), pp. 149–152
32. T. Kono, T. Ito, T. Tsuruda, T. Nishiyama, T. Nagasawa, T. Ogawa, Y. Kawashima, H. Hidaka, T. Yamauchi, in *IEEE ISSCC* (2013), pp. 212–214
33. Y. Taito, M. Nakano, H. Okimoto, D. Odaka, T. Ito, T. Kono, K. Noguchi, H. Hidaka, T. Yamauchi, in *IEEE ISSCC* (2015), pp. 132–134
34. D. Shum, J.R. Power, R. Ullmann, E. Suryaputra, K. Ho, J. Hsiao, C.H. Tan, W. Langheinrich, C. Bukethal, V. Pissors, G. Tempel, M. Rohrich, A. Gratz, A. Iserhagen, E.O. Andersen, S. Paprotta, W. Dickenscheid, R. Strenz, R. Duschl, T. Kern, C.T. Hsieh, in *IEEE IMW* (2012), pp. 139–142
35. S.T. Kang, B. Winstead, J. Yater, M. Suhail, G. Zhang, C.-M. Hong, H. Gasquet, D. Kolar, J. Shen, B. Min, K. Loiko, A. Hardell, E. Lepore, R. Parks, R. Syzdek, S. Williams, W. Malloch, G. Chindalore, Y. Chen, Y. Shao, L. Huajun, L. Louis, S. Chaw, in *IEEE IMW* (2012), pp. 131–134
36. G. Torrente, X. Federspiel, D. Rideau, F. Monsieur, C. Tavernier, J. Coignus, D. Roy, G. Ghibaudo, in *IEEE IRPS* (2016), pp. 5A-4
37. A. Baiano, M.V. Duuren, E.V. D. Vegt, B. Schippers, R. Beurze, D.T. Mofrad, H.V. Zwol, Y. Chen, J. Chiang, H. Lokker, K.V. Dijk, J. Verbree, Y.N. Chen, J. Garbe, R. Verhaar, D. Dormans, in *IEEE IMW* (2015), pp. 173–176
38. K. Ramkumar, I. Kouznesov, V. Prabhakar, K. Shakeri, X. Yu, Y. Yang, L. Hinh, S. Lee, S. Samanta, H. M. Shih, S. Geha, in *IEEE IMW* (2013), pp. 199–202
39. J. Chang, in *Leading Edge Embedded Non Volatile Memories* (2015)
40. C. Su, H. Tran, M. Tadayoni, N. Do, J. Yang, Method of making embedded memory device with silicon-on-insulator substrate, U.S. patent 9,431,407, 30 Aug 2016
41. N. Do, in *Leading Edge Embedded Non Volatile Memories* (2015)
42. V. Tiwari, in *Flash Memory Summit 2015—Driving Down the Memory Lane*
43. D. Fan, C. Chen, P. Tuntasood, Flash memory cells with separated self-aligned select and erase gates, and process of fabrication, U.S. patent 6,747,310, 8 June 2004
44. H. Tran, A. Ly, H. Nguyen, T. Vu, Array and pitch of non-volatile memory cells, U.S. patent 7,839,682, 23 Nov 2010
45. H. Tran, S. Nguyen, H. Nguyen, Sense amplifier for low voltage high speed sensing, U.S. patent 7,616,028, 10 Nov 2009

Chapter 6
SONOS 1Tr eFlash Memory

Hidenori Mitani and Ken Matsubara

6.1 History and Features of SONOS Technology

The SONOS flash memory using the CT structure has intrinsically higher reliability and affinity with logic-CMOS process than an FG-type cell and has been widely developed as a suitable cell for embedded applications. Especially, computerization of automotive control and development of IoT have expanded the application range, and three types cells are mainly developed according to the application as shown in Fig. 6.1. As a representative history of embedded SONOS flash memory, the world's first SG-type cell with 1.5Tr structure [1] was developed in the 1990s, and then TG-First SG-SONOS [2] was developed, furthermore, the world's first SG-SONOS with MG-zero-bias read scheme [3, 4] was developed as a suitable device for automotive applications. As for 1Tr structure, NROM [5] which realizes multi-bit scheme by using CT structure, and 1Tr-SONOS which overcomes the 1Tr-structure issues described in this chapter have been developed. A 2Tr structure [6, 7] has been developed as a low-cost integration in standard CMOS process. The 2Tr cell has larger sizes but can reduce overhead of the peripheral area to avoid over erase and deplete issues. Furthermore, the world's first Fin FET SONOS [8] and Twin MONOS [9] have also been developed for larger capacity and higher scaling.

Renesas Electronics has developed and productized SONOS technologies extensively as shown in Fig. 6.2. Based on 2Tr-SONOS technology for EEPROM stand-alone memory, two types of embedded SONOS technologies have been developed according to applications, 1Tr structure and 1.5Tr structure. 1Tr structure follows the low-power FN-tunneling principle which can be realized by thinning bottom film enabled by the intrinsic high reliability of the charge-trapping

H. Mitani (✉) · K. Matsubara
Core Technology Business Division, Renesas Electronics, 5-20-1, Josuihon-cho, Kodaira-shi, Tokyo 187-8588, Japan
e-mail: hidenori.mitani.zj@renesas.com

© Springer International Publishing AG 2018
H. Hidaka (ed.), *Embedded Flash Memory for Embedded Systems: Technology, Design for Sub-systems, and Innovations*, Integrated Circuits and Systems, DOI 10.1007/978-3-319-55306-1_6

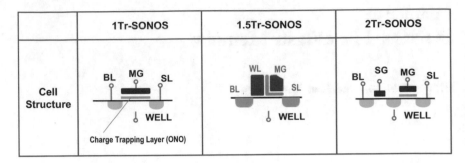

Fig. 6.1 SONOS-1Tr, 1.5Tr, and 2Tr-cell structures

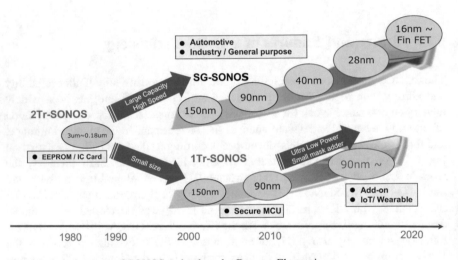

Fig. 6.2 Development of SONOS technology by Renesas Electronics

structure and eliminating the select transistor for high density. This structure has a good track record as a device suitable for the secure-MCU market, which requires low-power consumption and small area with 500 K endurance. By applying the mask reduction scheme through the elimination of the dedicated HV-MOS as well as the low-power consumption technology, the 1Tr structure can also be applied to an "add-on" market where low-cost integration of NVM is required and the first-ever 90 nm 1Tr-SONOS eFlash macros for automotive were reported in 2016 [10]. In addition, low-power consumption by the FN-tunnel rewrite also finds suitability in the IoT and wearable markets.

In contrast, the SONOS split-gate cell with 1.5Tr structure can be applied not only to general-purpose MCU but also to industry and automotive application MCUs because it has essentially high reliability due to the charge-trap structure as well as high performance and low-power read characteristics by adopting the split-gate structure. SONOS split-gate structure is especially suitable for automotive

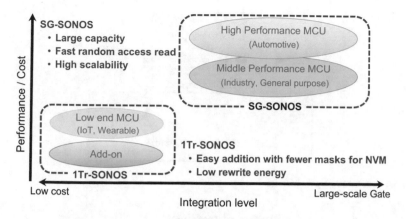

Fig. 6.3 Target applications of embedded SONOS technologies

application because high-speed random-access read characteristics, very low defect rate and harsh high-temperature durability are required. Renesas Electronics has adopted and mass-produced MCUs from 150 nm process generation for automotive, industrial and general purposes. After that, the first 40 nm SONOS split-gate eFlash macros for automotive use were developed in 2013 [3, 4] and the first 28 nm SONOS split-gate eFlash macros for automotive use were reported in 2015 [11, 12]. Furthermore, the first 16 nm FinFET SONOS split-gate was successfully fabricated and presented in 2016 [13]. Thus, SONOS split-gate memory cell shrinkage is steadily progressing thanks to its intrinsic scalability and simple structure.

In the target application map in Fig. 6.3, these embedded SONOS technologies provide a wide range of coverage in terms of integration, performance and cost.

6.2 1Tr-SONOS Cell Technology and Operations

1Tr-SONOS technology can be embedded with a minimal number of additional masks in base-line process. This technology features a simple one-transistor memory cell and can be programmed with low power due to FN-tunneling operation. In this section, basic cell operation and process technology of 1Tr-SONOS with fewer additional masks will be described.

Cell structure

Figure 6.4 shows the structure of a 1Tr-SONOS cell using ONO film ($SiO_2/Si_3N_4/SiO_2$) under the memory gate (poly-Si). This simple structure offers fewer additional masks and process steps than floating-gate (FG) structure. 1Tr cell structure has an intrinsic advantage compared with an FG cell in terms of high reliability because nitride film prevents the loss of overall charges from defect. Furthermore, the bottom-oxide thickness can be thinner than that of an FG cell

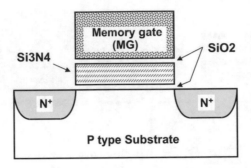

Fig. 6.4 Structure of 1Tr-SONOS

Table 6.1 Advantages and applications of 1Tr-SONOS

Features	Advantage	Application
Discrete charge-trapping	High reliability	Automotive, industrial
Simple 1Tr structure	Less additional cost	Add-on
FN tunnel rewrite	Low rewrite energy	IoT, Wearable

thanks to charge-trapping structure. Therefore, a SONOS cell can adopt FN-based program/erase operations of low current consumption.

Table 6.1 lists the advantages and target applications of 1Tr-SONOS technology to support consumer, industrial and automotive applications, in particular, for add-on uses. Because the number of additional masks is small, it can easily be adopted in products and processes in an "add-on" manner. These applications can now receive the benefits of embedded flash memory such as flexibility with respect to changing specifications, bug fixes, and storage and updating of parameters such as analog applications. In addition, because low-power program/erase operation is possible, it is also optimal for low-power applications such as IoT and wearable. However, a 1Tr-SONOS cell has intrinsic issues of over erase and read disturb, and solutions to these issues will be detailed in Sect. 6.3.

Program and erase operations

Figure 6.5 shows simplified program and erase operation conditions. Program and erase operations of 1Tr-SONOS adopt the FN-tunneling principle by using a high voltage (HV), which induces a high electric field at the bottom-oxide insulator to inject electrons and holes. Here, the typical value of HV required for the FN-tunneling phenomenon is 7 to 10 V, and thus it can be divided into middle voltage (MV [voltage approximately HV/2] over the memory gate (MV) and memory well (WELL). First, the program operation applies positive middle voltage (PMV) to the MG and negative middle voltage (NMV) to the WELL, source line (SL) and bit line (BL). The voltage difference between the MG and the WELL causes the FN-tunnel phenomenon with low-power consumption and injects electrons into the nitride-film layer. Next, in contrast, the erase operation applies negative middle voltage (NMV) to the MG and positive voltage to the WELL, the

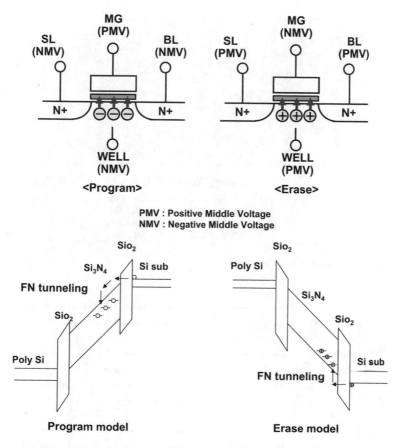

Fig. 6.5 Simplified cell operation of 1Tr-SONOS

SL and the BL. Holes are injected into the nitride-film layer by the FN tunneling. These FN-based operations realize low-power consumption compared with other eFlash technologies as shown in Fig. 6.6. In addition, 1Tr-SONOS realizes very narrow distributions in erased/programmed V_{th} without outlier bits as shown in Fig. 6.7.

Process technology

Figure 6.8 shows the details of general integration flows of 1Tr-SONOS memory cell. This device without a floating gate can realize a very simple flow and can be formed into various baseline process such as 1 Poly, 2 Poly, SONOS-first (Logic-last), SONOS-last (Logic-first), etc. The process of 1 Poly is the simplest process because the gate poly of the memory cell and the peripheral transistor is made common. However, this flow has an issue that is difficult to individually adjust the ONO film of the memory cell. Next, the 2 Poly process can solve the issue of the 1 Poly process by forming separately the gate poly of the memory cell and the

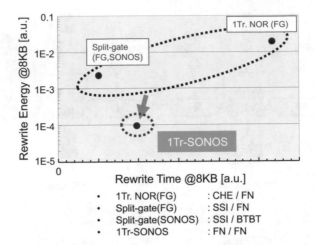

Fig. 6.6 Comparison of rewrite energy of eFlash

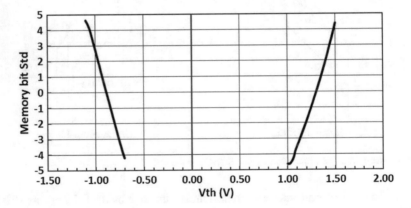

Fig. 6.7 Representative V_{th}-distribution characteristics of 1Tr-SONOS

peripheral transistor. Furthermore, the 2 Poly process is classified a two-type processe: SONOS-first (the right of Fig. 6.8) and SONOS-last (the middle of Fig. 6.8). SONOS-first is a process flow in the order of forming memory cells before forming peripheral transistors, in contrast, SONO-last is a process flow in the order of forming memory cells after forming peripheral transistors. Especially in SONOS-last process, it is necessary to consider the thermal-treatment impact to the peripheral transistor during memory-cell formation. Although selection of these processes should be made in consideration of affinity for the base process, 1Tr-SONOS can achieve lower cost than other eFlash in terms of additional mask number and process step. Therefore, 1Tr-SONOS cell is suitable for low-cost applications.

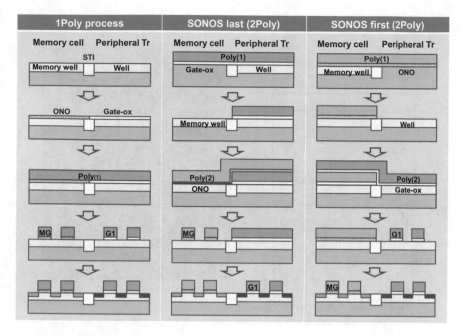

Fig. 6.8 1Tr-SONOS process flow

6.3   Basic Array Architecture and Operations

In this section, basic 1Tr-SONOS array architecture of Flash macro, which is composed of a memory array, a decoder, a sense circuit, control logic and power block are described together with their operations.

6.3.1   Array Architecture and Disturb Problems

Figure 6.9 and Table 6.2 show the architecture and micrograph of a 1Tr-SONOS 128 KB flash macro and key features. Flash macro is composed of a memory array, a row and column decoder, a sense amp, a mat isolation switch, control logic and a power circuit. This flash macro has two memory arrays of 64 KB, which are separated by a mat-isolation switch. Erase unit size is 2 KB, and the unit program page size per MG line is 128 B.

Figure 6.10 details operations of the 1Tr-SONOS array. The dashed region indicates the memory well. The voltages required for program and erase operations are distributed to the memory gate and the memory well. Conventional 1Tr-SONOS memory array suffers from issues of read disturb and program disturb. In the following sections, these disturb problems are identified and solutions are proposed.

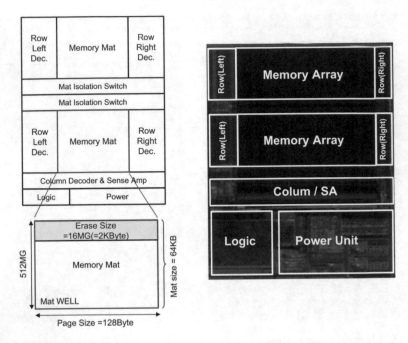

Fig. 6.9 1Tr-SONOS flash architecture and micrograph

Table 6.2 Key features of 1Tr-SONOS for automotive applications

Technology	90 nm 1Tr-SONOS
Metal layer	3 AL
Memory capacity	128 kB
Power supply	Core and I/O(VCC) 3.3 V ± 0.3 V
Operating temp. (T_j)	−40–175 °C
Read I/O number	32 bit + 7 bit
Random read freq.	52 MHz
Program time	3 ms/128 B
Erase time	5 ms/2 kB
P/E current	98 μA
P/E energy	0.07 mJ/8 kB
P/E endurance	>100 M cycles
Maximum capacity	512 kB

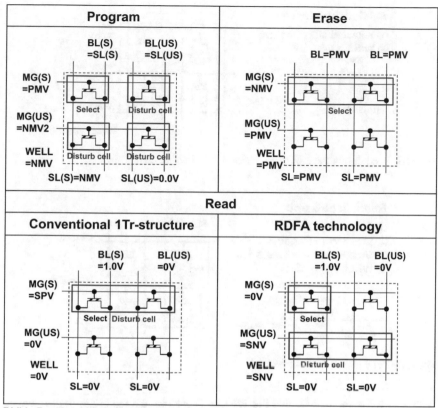

Fig. 6.10 1Tr-SONOS array operations and disturb problem

6.3.2 Read Operation

Conventional read operation

Figure 6.11 shows a read-disturb issue of a conventional 1Tr cell array. V_{th} of a conventional 1Tr-structure cell is set greater than 0 V to suppress channel leakage of unselected cells with 0 V on the MG in read operation. This is why a conventional 1Tr-cell array needs positive MG voltage in read operation, and the selected one suffers from read disturb because the voltage difference between MG and WELL is not 0 V. Even worse, to adopt FN tunneling for program and erase operations the bottom-oxide thickness should be thinner, which is critical for read disturb. In contrast, 1Tr-SONOS with RDFA technology can eliminate the voltage

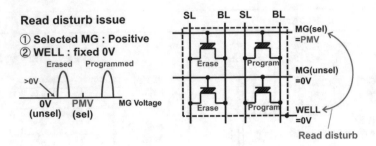

Fig. 6.11 Read-disturb issue of a conventional 1Tr-structure cell

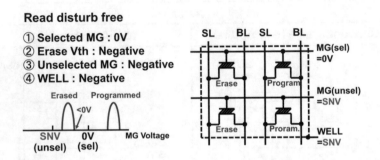

Fig. 6.12 Read disturb free architecture

difference between the MG and the WELL. Therefore, the read-disturb problem can be eliminated. Details are explained in following section.

Read disturb free architecture

Read disturb free architecture (RDFA) to solve the read-disturb issue of a 1Tr-cell array is shown in Fig. 6.12. The erased cell is depleted so that a MG = 0 V read operation can be performed. In addition, a negative SNV is supplied to the unselected MG and WELL, which sufficiently suppresses the leakage of unselected cells. With this architecture, the read-disturb issue can be solved for all three disturb cases as shown in Fig. 6.13.

(1) Selected erased cell: The voltage across ONO film is 0 V because a channel is formed.
(2) Selected programmed cell: Although selected programmed cells are in a very weak program condition (MG = 0 V, WELL = SNV), no actual problems occur because they are already programmed.
(3) Unselected cell: Read disturb in unselected cells is negligible because the voltage between the MG and the WELL, which is the dominant stressor in this case, is 0 V. Actually, read disturb hardly affects retention characteristics as shown in Fig. 6.14.

As a result, this array architecture realizes a read disturb free operation.

Cell State	Voltage across ONO film	Read Disturb
Selected cell (Erase)	0V	Disturb Free (MG=Channel=0V)
Selected cell (Program)	SNV	Weak disturb (Vth gain), but no problem for PRG cell.
Unselected cell	0V	Disturb Free (MG=WELL=SNV)

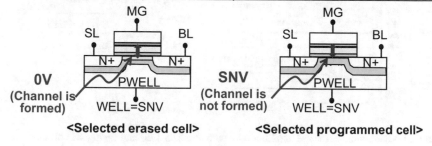

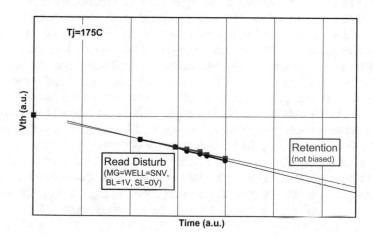

Fig. 6.13 Cell state of read disturb free architecture

Fig. 6.14 Read-disturb characteristics of a 1Tr-SONOS cell with RDFA

Read sense (charge-transfer sense amplifier with current-mirror load)

In general, read-cell current and cell size are in a trade-off condition, and various types of sense amplifier for high sensitivity are being studied. To solve this issue, a high-sensitivity charge-transfer sense amplifier was reported in 1975 [14]. The advantage of this scheme is that a low cell current is amplified to large voltage differences between the sense and the reference node, and high-speed sensing can be produced. In addition, various sense schemes based on charge-transfer have been reported including a cross-coupled charge-transfer sense amplifier in 1975 [15] being one of the schemes.

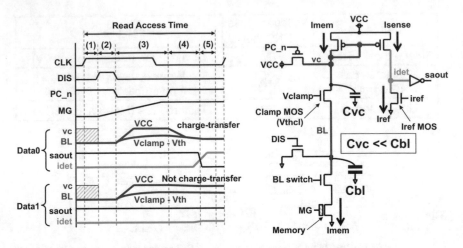

Fig. 6.15 Charge-transfer sense amplifier with current-mirror load and timing waveforms

In 1Tr-SONOS, current-mirror load type based on charge-transfer is used, and this scheme is described in this section. Figure 6.15 shows a charge-transfer sensing scheme with current-mirror load. Read-access time with this scheme consists of five parts; (1) address input and decoding, (2) BL discharge, (3) MG activation and BL pre-charge, (4) data sensing, and (5) logic-data output.

The pre-charge operation is carried out through the clamp MOS when the BL voltage is set to $(V_{clamp} - V_{thcl})$ level. This method through the clamp MOS prevents read disturb because the BL voltage is limited under several voltages (e.g., 1 V). After the pre-charge operation, because the BL voltage is slightly decreased by the cell current (I_{mem}), the node (vc) that was closed by the clamp MOS is greatly amplified by charge-transfer because C_{vc} is considerably smaller than C_{bl}, and the clamp MOS works the source follower under conditions of minute changes in BL voltage. After vc is amplified, the detected current $(I_{sense} = I_{mem})$ flows by the current mirror, and I_{sense} is compared with the reference current (I_{ref}). The node (idet) is amplified by the difference between I_{sense} and I_{ref}, and the reading data is output as a sense amplifier-output signal (saout).

Comparison of the latch type [14] and current-mirror load type is shown in Table 6.3. The access speed of the latch type is faster than that of the current-mirror type because the sense and reference (ref) node are amplified by latch; therefore, sense timing (saenb) is critical because it is susceptible to device variations. In contrast, current-mirror load provides timing-free operation and can be read by single-end.

Table 6.3 Comparison sense scheme

Sense scheme based on charge-transfer	Current-mirror load	Latch type
Access speed	Middle-speed	High-speed
Sense timing	Timing-free	Should be optimized
Circuit element		

6.3.3 Erase and Program

Erase operation

Figure 6.16 shows the details of erase operation. Erase voltage is applied to the MG of the selected page and the WELL. The voltage across the MG and the WELL causes FN-tunneling erase. As for unselected page, by applying the same voltage to the WELL and the MG, the unselected pages are free from erase disturb.

Next, the deplete issue in erase operation by the continuous erase pulse will be discussed. In the case of a 1Tr-structure cell, the continuous-erase pulse causes an undesirable decrease of the V_{th} level as shown in Fig. 6.17. To solve this issue, 1Tr-SONOS with FN programming adopts a pre-write scheme.

Pre-write operation

This section describes the pre-write operation. A 2Tr-structure cell is free from depletion issue because the select transistor suppresses channel leakage even with a depleted erased cell. However, a 1Tr-structure cell without a select transistor can not cut off the leakage of the unselected erased cells. Therefore, the continuous-erase pulse to erased cells can cause a gradual decrease of the V_{th} level as shown in Fig. 6.17 (1). To avoid this issue, a weak program pulse is applied before applying the erase pulse. This weak program pulse is called "pre-write". This pulse can suppress undesirable decrease of the V_{th} level of erased cells without affecting the V_{th} level of programmed cells and can also control the erased worst V_{th} level [Fig. 6.17 (2)]. In addition, the pre-write pulse duration is negligible compared with the erase pulse time and does not affect the specification. The measured characteristics are shown in Fig. 6.18.

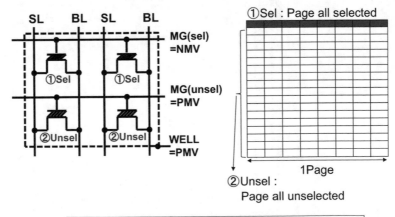

No.	Cell State	Cell mode
①	Sel	Erase (FN tunneling)
②	Unsel	Disturb free

Fig. 6.16 Erase operation and disturb mode

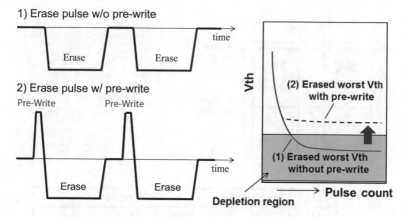

Fig. 6.17 Pulse sequence of pre-write operation and V_{th} distribution

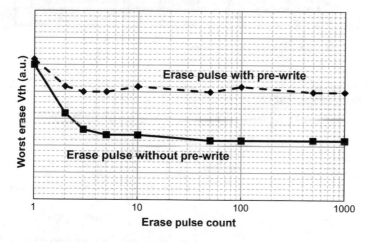

Fig. 6.18 Measured pre-write characteristics

Program operation and disturb

Figure 6.19 shows the details of program operation and disturb mode. Program voltage is applied to the selected MG and the WELL. The voltage difference between the MG and the WELL can cause FN tunneling. In contrast, three disturb modes exist in unselected cells. Each mode is as follows.

1. Disturb inhibit: a stress of the unselected cells on the selected MG.
2. Disturb gain: a stress of the selected cells on the unselected MG.
3. Disturb loss: a stress of the unselected cells on the unselected MG.

Disturb inhibit is generated by a voltage difference between selected the MG and unselected the SL; disturb loss is generated by a voltage difference between unselected SL and the unselected MG; and disturb gain is generated by a voltage

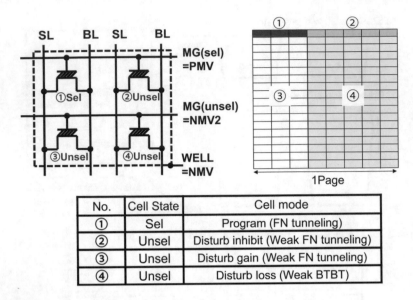

Fig. 6.19 Program operation and disturb mode

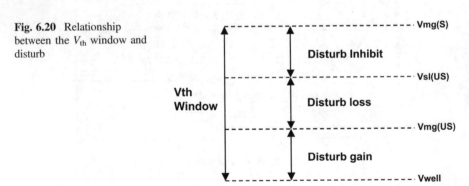

Fig. 6.20 Relationship between the V_{th} window and disturb

difference between the unselected MG and WELL. Disturb inhibit and disturb gain are caused by weak FN tunneling, and disturb loss is caused by weak BTBT (band-to-band-tunneling).

The most important point in designing the program operation is to assure that the V_{th} window overcomes data loss in the disturb mode. The initial V_{th} window, as determined by the voltage across selected MG ($V_{mg}(S)$) and WELL (V_{well}), should be larger than all the disturb effects summed together as shown in Fig. 6.20.

6.3.4 Modified MG Decoder

One major target for 1Tr-SONOS technology is to minimize the additional mask count. To cause FN tunneling when programming and erasing memory cells, high

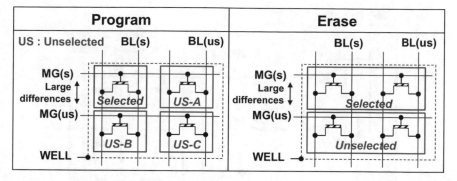

Fig. 6.21 1Tr-SONOS array architecture

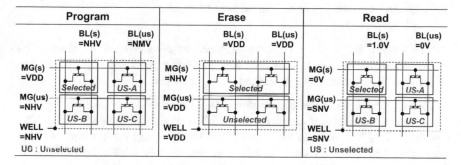

Fig. 6.22 1Tr-SONOS-array architecture using a conventional MG decoder

voltages are applied to ONO film in 1Tr-SONOS such that there are large voltage differences between the selected memory gate (MG(s)) and the unselected memory gate (MG(us)). This is done because voltage applied to the MG(us) should be approximately equal to the WELL voltage to prevent a disturb issue as shown in Sect. 6.3.3 and Fig. 6.21. These voltage differences can not be coped with by standard IO-MOS, so the tolerance of high breakdown-voltage is generally required for the peripheral transistors to control the MG. If the base process does not have high breakdown-voltage transistors, it is necessary to add a few masks for implementing such transistors, or circuit techniques without high breakdown-voltage transistors should be adopted.

In the following section, the basic structures of a conventional MG decoder with high breakdown-voltage transistors are shown, and a modified MG decoder without high breakdown-voltage transistors is proposed.

Basic structure and operation of the conventional MG decoder
The architecture, using as conventional MG decoder shown in Fig. 6.22, employs high breakdown-voltage transistors. "V_{DD}" stands for core-logic voltage; "NHV" stands for negative high voltage; "NMV" stands for negative middle voltage; and

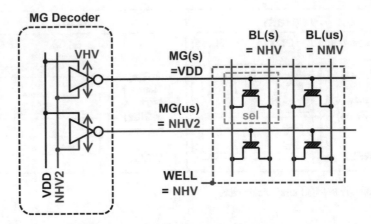

Fig. 6.23 Program conditions of the conventional MG decoder

"SNV" stands for slightly negative voltage. In conventional architecture, an internal-voltage generator creates only negative voltages, which are applied to the MG or the WELL according to each mode of operation.

Program conditions of the conventional MG decoder

The program conditions of the conventional MG decoder are shown in Fig. 6.23. In this mode, V_{DD} is applied to the selected MG (MG(s)), and NHV is applied to the unselected MG (MG(us)) and WELL, thus resulting in alleviating the program disturb between the MG(us) and the WELL. VHV means "large voltage differences" (e.g., >10 V) between V_{DD} and NHV2. A thick oxide film is required to control these high voltages, which means that VHV can not be addressed by standard IO-MOS. In other words, the peripheral transistors must have high breakdown-voltage, and additional masks to control these high voltages are necessary if the base process does not provide control.

Erase conditions of the conventional MG decoder

The erase conditions of the conventional MG decoder are shown in Fig. 6.24. In this mode, NHV is applied to the MG(s), and V_{DD} is applied to the MG(us) and the WELL, thus preventing erase disturb between the MG(us) and the WELL. In addition, the peripheral transistors must also have high breakdown-voltage, and additional masks to control these high voltages are necessary as the same under program conditions.

Read conditions of the conventional MG decoder

The read conditions of the conventional MG decoder are shown in Fig. 6.25. This configuration is based on RDFA architecture as shown in Sect. 6.3.2. In this mode, 0 V is applied to the MG(s), and SNV is applied to the MG(us) and the WELL, thus preventing read disturb between the MG(us) and the WELL. These voltages applied to the MG can be controlled by standard IO-MOS, so additional masks to control these voltages are not necessary.

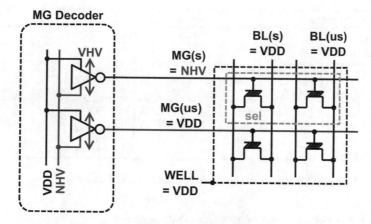

Fig. 6.24 Erase conditions of the conventional MG decoder

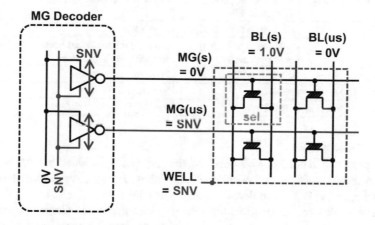

Fig. 6.25 Read conditions of the conventional MG decoder

Modified MG decoder for relaxing high voltages

This section shows a modified MG decoder eliminating high breakdown-voltage transistors composed of standard IO-MOS for lower cost. Figure 6.26 shows the array architecture by using a modified MG decoder. "V_{DD}" stands for core logic voltage; "PMV" stands for negative middle voltage; "PMV" stands for positive middle voltage; and "SNV" stands for slightly negative voltage.

In the modified architecture, high voltages for programming and erasing memory cells are divided into PMV and NMV. Because these voltages are not so high (e.g., 5 V), this architecture can be controlled by the decoder without high breakdown-voltage transistors.

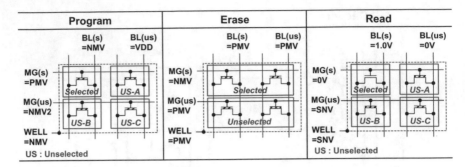

Fig. 6.26 1Tr-SONOS array architecture using a modified MG decoder

Table 6.4 Function assignment of the modified MG decoder

		MG decoder left	MG decoder right
Function assignment		Read and negative voltage in P/E	Positive voltage in P/E
Speed		High-speed (for read)	Low-speed
Area		Optimized to high-speed read operation	Small (minimum size) because high-speed operation isn't necessary
MG drive voltage	Read	0 V(s)/SNV(us)	Not used
	Erase	NMV(s)	PMV(us)
	Program	NMV2(us)	PMV(s)

Function assignment

In the proposed modified MG decoder, the left and right sides are dedicated to negative middle voltages and positive middle voltages, respectively, thus alleviating high voltages (VHV) at the decoder devices as shown in Table 6.4.

The left-side decoder controls the read state and negative voltages in programming and erasing (P/E). In contrast, the right-side decoder controls only positive voltages in P/E. The left-side decoder is optimized to operate high-speed at read mode, but the right-side decoder is not. Therefore the area of the right-side decoder is minimal, thus alleviating area penalty by two-side decoder structure.

Program conditions of the modified MG decoder

The structure and program conditions of the modified MG decoder are shown in Fig. 6.27. VIO is the middle voltage differences (e.g., <5 V) between V_{DD} and NMV2, and these voltages are applied to the oxide film of the transistors, They are not large, so the standard IO-MOS can control them. In this mode, PMV is applied to the selected MG (MG(s)), and NMV (NMV2) is applied to the unselected MG (MG(us)) and WELL, thus resulting in alleviating the program-disturb issue between the MG(us) and the WELL.

In this mode, voltages applied to the MG are controlled from the left and right sides. The left-side decoder controls negative voltage (NMV2), and the right-side decoder controls positive voltage (PMV). Each side decoder is divided by transfer

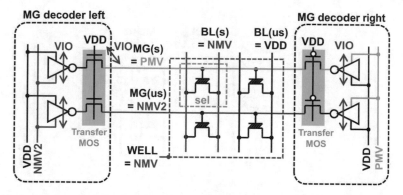

Fig. 6.27 Program conditions of the modified MG decoder

MOS, and high voltages applied to decoder transistors are relaxed under VIO that can be coped with by standard IO-MOS. As a result, this structure can be configured without using high breakdown-voltage transistors but only using standard IO-MOS.

Erase conditions of the modified MG decoder

The structure and erase conditions of the modified MG decoder are shown in Fig. 6.28. In this mode, NMV is applied to the MG(s), and PMV is applied to the MG (us) and the WELL, thus preventing erase disturb between the MG(us) and the WELL.

In this mode, voltages applied to the MG are controlled from the left and right sides. The function of each side decoder is same as the program conditions, and high voltages applied to transistors are relaxed under VIO thanks to transfer MOS. Therefore, this structure can be composed of standard IO-MOS.

Read conditions of the modified MG decoder

Read conditions of the modified MG decoder are shown in Fig. 6.29. This configuration is based on RDFA architecture as shown in Sect. 6.3.2. In this mode, 0 V

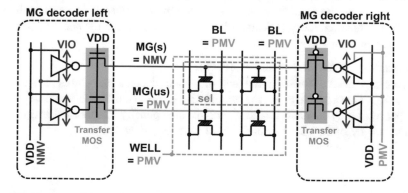

Fig. 6.28 Erase conditions of the modified MG decoder

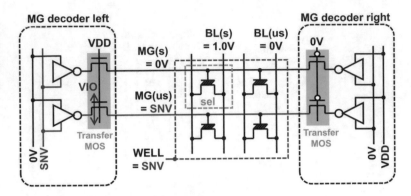

Fig. 6.29 Read conditions of the modified MG decoder

Table 6.5 Comparison of the MG-decoder structure

	Modified	Conventional
Decoder configuration	Two side decoder	One side decoder
Required breakdown-voltage	VIO (VIO can be coped with by standard IO-MOS)	VHV (high voltage)
High breakdown-voltage transistors	Not used (only standard IO-MOS are used)	Necessary (e.g.) 5V-IO
Area	Penalty is minimal (Same size as that of conventional 1)	Ref

is applied to the MG(s), and SNV is applied to the MG(us) and the WELL, thus preventing read disturb between the MG(us) and the WELL.

In this mode, voltages applied to the MG are controlled from only left side. SNV are not so large and thus can be controlled by standard IO-MOS. In terms of operating speed and area, the MG decoder of left side should be optimized to operate at high-speed for read operation. In contrast, the MG decoder of right side is composed of minimum-size devices for minimum area penalty. As a result, the area penalty by modified the MG decoder is minimal even though the decoder is divided into left and right sides.

Comparison of the MG-decoder structures
Comparison of the conventional and modified decoder structures is shown in Table 6.5. The advantage of the modified decoder is that the structure can be composed of only standard IO-MOS, and the area penalty by modified MG decoder is minimal by optimizing the function of each side decoder.

6.3.5   Adaptable Slope Pulse Control (ASPC) Technique

When programming and erasing memory cells, the ONO film is damaged by FN-stress, thus degrading reliability. This section describes a technique—called "adaptable slope pulse control (ASPC) technique"—to solve this problem.

Conventional Step Pulse Control technique

Step pulse control technique was conventionally adopted to alleviate FN-stress. Figure 6.30 shows a conventional high-voltage (V_{out}) generation sub-system with step pulse control technique and FN-stress calculations by T-CAD simulation. Here, "V_{out}" means negative voltage applied to the MG.

In this technique, the steep change of FN-stress on the ONO film is severe in each step transition due to steep output-voltage (V_{out}) change in automotive usage. In addition, the ring oscillator (OSC) with fixed-clock frequency causes large variation of output current (I_{out}) under the whole process/voltage/temperature conditions, leading to wide and high voltage (V_{out}) slope variation. To alleviate the worst case FN-stress, it is effective to keep I_{out} constant under the whole conditions.

Adaptable Slope Pulse Control (ASPC) technique

ASPC technique can solve these problems. Figure 6.31 shows an advanced high-voltage (V_{out}) generation sub-system with ASPC technique and FN-stress calculation value by T-CAD simulation. The ASPC block diagram includes an intelligent slope pulse control circuit (ISPCC), a smart-clock generator (SCG) and a voltage detector (VD).

The V_{out} is generated with a smooth slope pulse by ISPCC; as a result, the steep change of FN-stress on the ONO film can be alleviated sufficiently. Furthermore, the SCG adjusts the charge-pump clock frequency (f_{clk}) by detecting a signal (idet)

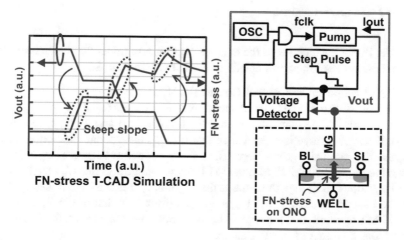

Fig. 6.30 V_{out} generator sub-system with step pulse control and FN-stress calculations by T-CAD simulation

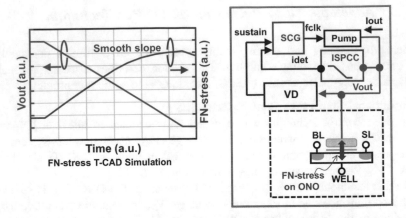

Fig. 6.31 V_{out}-generator sub-system with ASPC and FN-stress calculation values by T-CAD simulation

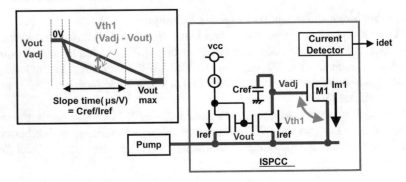

Fig. 6.32 The details of ISPCC

from the ISPCC, which keeps the average I_{out} constant under the whole conditions. The following section shows details of the ISPCC and the SCG.

Intelligent Slope Pulse Control Circuit (ISPCC)

The details of the ISPCC circuit are shown in Fig. 6.32. The V_{adj} slope in the ISPCC circuit is determined by a reference current (I_{ref}) and a reference capacitor (C_{ref}). If the V_{out} slope is steeper than that of V_{adj}, the voltage difference between V_{adj} and V_{out} gradually grows large. Then, if the difference becomes larger than V_{th1} that is the V_{th} of the NMOS M1, the M1 turns on, and the current I_{m1} starts to flow. The V_{out} slope can be kept the same as the V_{adj} slope, which is determined by the I_{ref} and the C_{ref}. In contrast, if the V_{out} slope is slower than that of the V_{adj} slope, the NMOS M1 turns off, and the current I_{m1} does not flow. As a result, the V_{out} slope is not steeper than that of the V_{adj} slope.

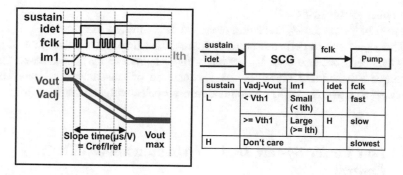

Fig. 6.33 The details of SCG

However, a residual current is thrown away as I_{m1} to generate the V_{out} slope same as the V_{adj} slope at this rate, which leads to ineffective power consumption. To solve this problem, the current detector in the ISPCC circuit monitors I_{m1} and sends the detect signal (idet) to SCG, and the SCG adjusts f_{clk} (charge-pump clock frequency) dynamically.

Smart Clock Generator (SCG)

SCG adjusts the charge-pump clock frequency (f_{clk}) to keep I_{out} constant and obtain smooth V_{out} slope with much less variation as shown in Fig. 6.33. If the idet signal shows "L" level, which means that V_{out} slope is slower than that of V_{adj}, SCG adjusts f_{clk} fast. In contrast, if the idet signal shows "H" level, which means that V_{out} slope is steeper than that of V_{adj}, SCG adjusts f_{clk} slow. In this way, SCG adjusts f_{clk} dynamically and generates constant I_{out} under the whole condition, which helps the V_{out} slope constantly without residual power consumption.

In addition, once the VD detects that V_{out} has reached target voltage (V_{outmax}), the whole system transits to "sustain mode." When the system transits to sustain mode once, f_{clk} is set the slowest to just sustain V_{outmax} level, resulting in low power consumption. Consequently, the ASPC technique realizes high-voltage generation with less FN-stress on the ONO film as well as effective power consumption. Actually, total P/E current can be reduced to only 98 µA.

Fig. 6.34 Endurance characteristics

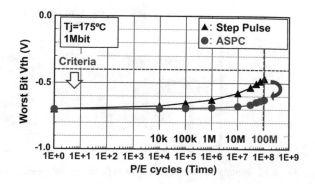

The effect of the ASPC

Figure 6.34 shows the V_{th} transition of erased cells during P/E cycles at T_j of 175 ° C. The V_{th} of erased cells are more critical against read criteria than that of programmed cells because both V_{th} gradually increase as P/E cycles increase. ASPC technique substantially improves the V_{th} increase of erased cells by alleviating FN-stress and realizes >100 million endurance cycles with sufficient margin.

6.4   Low Power System Design by Autonomous eFlash Control

Battery-power consumption has been increasing due to the increase in the number of mounted semiconductor devices in automotive applications. Especially, the spread of idling stop technology and anti-theft devices, the battery consumption becomes a problem during engine stop as shown in Fig. 6.35. In this section, advanced technology of 1Tr-SONOS based low-power system will be described.

Conventional system

"Green" cars are adopting engine idle-reduction technology (EIRT) for better fuel efficiency. With EIRT, when a vehicle is stationary the engine stops automatically, and the battery is the sole power supply to all electronic devices. Even during this period, MCUs in each motor-control system may work and store calibration data to the NVM. Therefore, green cars require efficient MCU-power management. Figure 6.36 shows a conventional analog-control MCU-system block diagram and its operation. Battery-power consumption has been increasing due to an increase of the number of mounted semiconductor devices. Especially, with the spread of idling-stop technology and anti-theft devices, battery consumption becomes a problem during engine stop. In this situation, reduction of power consumption at the system level is strongly required. However, the CPU of a conventional system must be always active to control external EEPROM during engine stop. Moreover, the P/E current of an external EEPROM is large.

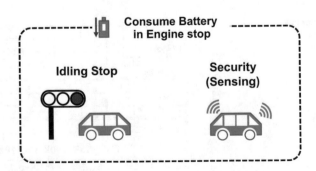

Fig. 6.35 Battery-consumption situation in automotive applications

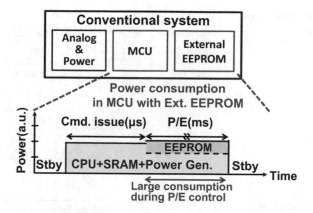

Fig. 6.36 Conventional analog-control system

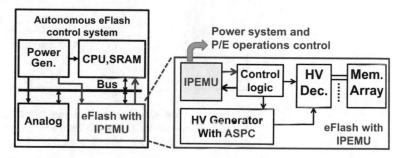

Fig. 6.37 Low-power system under autonomous eFlash control

Autonomous eFlash-control system

Figure 6.37 shows a low-power MCU system with 1Tr-SONOS macro including an idling P/E-management unit (IPEMU). The IPEMU is designed to manage eFlash and system-operation modes instead of the CPU, thus the CPU and SRAM-activity factor is smaller. In addition, data storage-power consumption is drastically reduced because the 1Tr-SONOS eFlash macro requires only 98 μA for P/E operations, which is much less than required by the external EEPROM ($\sim$5 mA). Moreover, the stand-by V_{dd} generator can sustain V_{dd} during P/E operations thanks to the small P/E current, which results in less current consumption by the power unit. In this way, autonomous control of the eFlash-power system and P/E operations greatly reduces power consumption. The details of IPEMU will be described in the following section.

Idling P/E-management unit (IPEMU)

Figure 6.38 shows the control of a proposed 1Tr-SONOS eFlash system with an IPEMU. After waking up from stand-by status, the CPU issues commands to the 1Tr-SONOS eFlash macro to store data. After the CPU issues commands, it reverts

Fig. 6.38 Control wave form
of IPEMU scheme

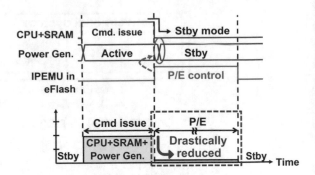

Fig. 6.39 Comparison of
power consumption

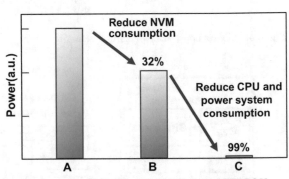

A : Conv. System on board with External EEPROM
B : Monolithic chip by 1Tr-SONOS w/o IPEMU
C : Monolithic chip by 1Tr-SONOS w/i IPEMU

to standby-mode, and the IPEMU controls the power generator and eFlash P/E
operation. As for the power generator, its capability can be reduced from active V_{dd}
to standby V_{dd} during P/E operation thanks to the small P/E current of the proposed
1Tr-SONOS macro. As a result, the proposed system can minimize the CPU/SRAM
active period and achieve drastically reduced system power. The
1Tr-SONOS-based MCU system reduces current consumption by 99% compared
with a conventional one as shown in Fig. 6.39.

6.5 Conclusion

A 1Tr-SONOS eFlash-memory cell with charge-trapping storage offers the following attractive features: simple structure results in fewer additional masks and
fewer process steps. The charge-trapping structure has an intrinsic advantage
compared with an FG cell in term of high reliability. FN-based program and erase
operation (thanks to the charge-trapping structure) can realize low current consumption. In this way, 1Tr-SONOS technology can support consumer, industrial

and automotive applications. Especially, because of its low-cost features, it can easily be added in products and processes in which flash memory was previously not embedded.

A 128 kB 1Tr-SONOS eFlash macro introduced in this chapter was designed and fabricated in a 90 nm 1Tr-SONOS eFlash process. A 98 μA P/E current is achieved due to the simple cell and array structure, and the adaptable slope pulse control (ASPC) technique. P/E energy is 0.07 mJ/8 kB at a T_j of 175 °C and almost equivalent to that of a ReRAM macro for consumer use [16]. Moreover, the 100M-cycle endurance and P/E energy of 0.07 mJ/8 kB at T_j of 175 °C will aid in expanding eFlash into automotive applications.

References

1. W.M. Chen, C. Swift, D. Roberts, K. Forbes, J. Higman, B. Maiti, W. Paulson, K.T. Chang, A novel flash memory device with split gate source side injection and ONO charge storage stack (SPIN). in *Symposium on VLSI Technology*, pp. 63–64 (1997)
2. F. Ito, Y. Kawashima, T. Sakai, Y. Kanamaru, Y. Ishii, M. Mizuno, T. Hashimoto, T. Ishimaru, T. Mine, N. Matsuzaki, H. Kume, T. Tanaka, Y. Shinagawa, T. Toya, K. Okuyama, K. Kuroda, K. Kubota, A novel MNOS technology using gate hole injection in erase operation for embedded nonvolatile memory applications. in *Symposium on VLSI Technology*, pp. 80–81 (2004)
3. T. Kono, T. Ito, T. Tsuruda, T. Nishiyama, T. Nagasawa, T. Ogawa, Y. Kawashima, H. Hidaka, T. Yamauchi, 40 nm embedded SG-MONOS flash macros for automotive with 160 MHz random access for code and endurance over 10 M cycles for data. in *ISSCC Digest of Technical Papers*, pp. 212–214 (2013)
4. T. Kono, T. Ito, T. Tsuruda, T. Nishiyama, T. Nagasawa, T. Ogawa, Y. Kawashima, H. Hidaka, T. Yamauchi, 40-nm embedded split-gate MONOS (SG-MONOS) flash macros for automotive with 160-MHz random access for code and endurance over 10 M cycles for data at the junction temperature of 170 °C. IEEE J. Solid-State Circuits **49**(1), 154–166 (2014)
5. B. Eitan, P. Pavan, I. Bloom, E. Aloni, A. Frommer, D. Finzi, Can NROM, a 2-bit, trapping storage NVN cell, give a real challenge to floating date cells. in *International Conference on Solid State Devices and Materials*, pp. 522–524 (1999)
6. H.M. Lee, S.T. Woo, H.M. Chen, R. Shen, C.D. Wang, L.C. Hsia, C.C.-H. Hsu, NeoFlash®—true logic single poly flash memory technology. in *IEEE Non-Volatile Semiconductor Memory Workshop*, pp. 15–16 (2006)
7. K. Ramkumar, I. Kouznetsov, V. Prabhakar, K. Shakeri, X. Yu, Y. Yang, L. Hinh, S. Lee, S. Samanta, H.M. Shih, S. Geha, P.C. Shih, C.C. Huang, H.C. Lee, S.H. Wu, J.H. Gau, Y.K. Sheu, A scalable, low voltage, low cost SONOS memory technology for embedded NVM applications. in *IMW*, pp. 119–202 (2013)
8. C.W. Oh, S.D. Suk, Y.K. Lee, S.K. Sung, J.-D. Choe, S.-Y. Lee, D.U. Choi, K.H. Yeo, M.S. Kim, S.-M. Kim, M. Li, S.H. Kim, E.-J. Yoon, D.-W. Kim, D. Park, K. Kim, B.-I. Ryu, Damascene gate FinFET SONOS memory implemented on bulk silicon wafer. in *IEEE International Electron Devices Meeting*, pp. 893–896 (2004)
9. Y. Hayashi, S. Ogura, T. Saito, T. Ogura, Twin MONOS cell with dual control gates. in *Symposium on VLSI Technology*, pp. 122–123 (2000)
10. H. Mitani, K. Matsubara, H. Yoshida, T. Hashimoto, H. Yamakoshi, S. Abe, T. Kono, Y. Taito, T. Ito, T. Kurafuji, K. Noguchi, H. Hidaka, T. Yamauchi, A 90 nm embedded 1T-MONOS flash macro for automotive applications with 0.07 mJ/8 kB rewrite energy and

endurance over 100 M cycles under T_j of 175 °C. in *IEEE Solid-State Circuits Conference, Digest of Technical Papers*, pp. 140–141 (2016)

11. Y. Taito, M. Nakano, H. Okimoto, D. Okada, T. Ito, T. Kono, K. Noguchi, H. Hidaka, T. Yamauchi, A 28 nm embedded SG-MONOS flash macro for automotive achieving 200 MHz read operation and 2.0 MB/s write throughput at T_j of 170 °C. in *ISSCC Digest of Technical Papers*, pp. 132–133 (2015)

12. Y. Taito, T. Kono, M. Nakano, T. Saito, T. Ito, K. Noguchi, H. Hidaka, T. Yamauchi, A 28 nm embedded split-gate MONOS (SG-MONOS) flash macro for automotive achieving 6.4 GB/s read throughput by 200 MHz no-wait read operation and 2.0 MB/s write throughput at T_j of 170 °C. IEEE J. Solid-State Circuits **51**(1), 213–221 (2016)

13. T. Hagiwara, Y. Yatsuda, R. Kondo, S. Minami, T. Aoto, Y. Itoh, A 16 kbit electrically erasable PROM using n-Channel Si-Gate MNOS technology. IEEE J. Solid-State Circuits **SC-15**(3), 346–353 (1980)

14. L.G. Heller, D.P. Spampinato, Y.L. Yao, High-sensitivity charge-transfer sense amplifier. in *IEEE Solid-State Circuits Conference, Digest of Technical Papers*, pp. 112–113 (1975)

15. L.G. Heller, Cross-coupled charge-transfer sense amplifier, in *IEEE Solid-State Circuits Conference, Digest of Technical Papers*, pp. 20–21 (1979)

16. M. Ueki, K. Takeuchi, T. Yamamoto, A. Tanabe, N. Ikarashi, M. Saitoh, T. Nagumo, H. Sunamura, M. Narihiro, K. Uejima, K. Masuzaki, N. Furutake, S. Saito, Y. Yabe, A. Mitsuiki, K. Takeda, T. Hase, Y. Hayashi, Low-power embedded ReRAM technology for IoT applications, in *IEEE Symposium on VLSI Circuits*, pp. 108–109 (2015)

Chapter 7
SONOS Split-Gate eFlash Memory

Takashi Ito and Yasuhiko Taito

7.1 Basic Cell Technology and Operations

The SONOS split-gate cell can be applied not only to general-purpose MCU but also to industry and automotive application MCUs because it has essentially high reliability due to its charge-trap structure, high-performance and low-power read characteristic by adopting the split-gate structure. The SONOS split-gate structure is especially suitable for automotive application because high-speed random-access read characteristics, a very low defect rate, and harsh-high temperature durability are required.

The advantages of SONOS split-gate eFlash in the MCU applications are shown in Fig. 7.1. Intrinsically, charge trapping (CT)-type storage structures are advantageous due to their high reliability and scalability. In addition, thin nitride film storage realizes a low-profile cell and enables high compatibility with base-logic CMOS. Split-gate or 1.5Tr structure enables low power programming by using source-side injection channel hot-electron (SSI-CHE). Moreover, a low-voltage word line driver and no high voltage (HV) in the read-access path realize high-speed random-read >100 MHz, lower-power read operation, and small flash macro area.

Figure 7.2 shows a comparison between an NOR-type floating-gate (FG) cell and a SONOS CT-type cell. In terms of the FG-type cell, the storage area is the conductor, and the storage node is a single node. If one conductive defect exists in the oxide film around the FG, all the stored charges are finally lost through the defect. In contrast, regarding the CT-type cell, the storage area is the charge trap in

T. Ito (✉) · Y. Taito
Core Technology Business Division, Renesas Electronics, 5-20-1, Josuihon-cho, Kodaira-shi, Tokyo 187-8588, Japan
e-mail: takashi.ito.xt@renesas.com

Y. Taito
e-mail: yasuhiko.taito.zc@renesas.com

© Springer International Publishing AG 2018
H. Hidaka (ed.), *Embedded Flash Memory for Embedded Systems: Technology, Design for Sub-systems, and Innovations*, Integrated Circuits and Systems, DOI 10.1007/978-3-319-55306-1_7

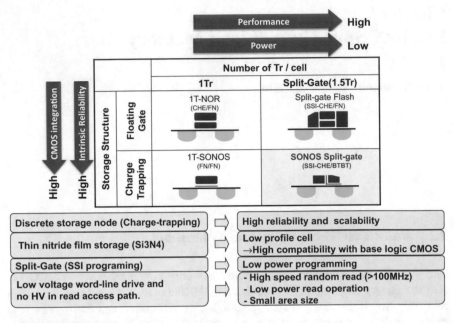

Fig. 7.1 SONOS split-gate eFlash memory

	NOR type (Floating gate)	SONOS: Charge-trapping type
Device structure		
Storage area	Floating gate (conductor)	Charge trap in SiN film (insulator)
Storage node	Single node	Multi node (discrete)
Influence of defect	All of the stored charges are lost through the defect	Only limited charges located near the defect are lost
Integration with CMOS Process	Will see difficulties (due to high cell profile)	Very Good (by low cell profile)

Fig. 7.2 Advantages of a charge trapping-type cell

thin silicon-nitride film with an insulator, and the storage node is the multi node. Thanks to these separately located multi traps, only the charges near the defect are lost. That is, a CT-type cell is intrinsically more reliable than an FG-type cell. In addition, regarding integration with the CMOS process, the gate height of logic

CMOS transistors have been getting lower along with scaling, which makes it difficult to integrate certain types of eFlash into advanced-logic CMOS process due to high cell profile. In contrast, low-profile eFlash cell shows a high affinity with advanced-logic CMOS process. This means that the cell profile can be a decisive factor in the integration of eFlash-memory cells into logic CMOS process in advanced technology node. Affinity with logic CMOS process has become much more important than ever, especially beyond 40 nm. From this viewpoint, SONOS CT-type eFlash-memory cells have great advantages over FG-type eFlash-memory cells since the cell height of a CT-type cell is lower than that of an FG-type cell thanks to thin film storage. A CT-type cell can be integrated within the same gate height as logic transistors, even in advanced technology nodes. This means that a CT-type cell has much better scalability than an FG-type cell from the viewpoints of reliability and affinity with the advance of logic CMOS process.

Split-gate (1.5Tr) cells have two advantages over 1Tr cells (Fig. 7.3) in terms of performance: (1) very efficient programming due to source-side injection (SSI) capability; and (2) no drain-leakage current by series connection of a select transistor. Especially, the second advantage allows a negative threshold-voltage (V_{th}) setting of erased cells. This means that there is no need to boost the WL in read operation, and the fast-read path can be built only with fast low-voltage logic CMOS transistors.

A SONOS split-gate cell with the combination of split-gate and charge-trapping using a nitride-film (Si_3N_4) structure is shown in Figs. 7.4 and 7.5. As described previously, a charge-trapping structure with Si_3N_4 offers intrinsically high data reliability because charge storage in the separately located traps in the nitride film prevents a defect from draining the overall charges. From the viewpoint of poly structure, the SONOS split-gate cell is effective for cell scaling and low bit-failure rate because it has simple structure with only two poly nodes. This feature promises sufficient charge storage in a smaller cell.

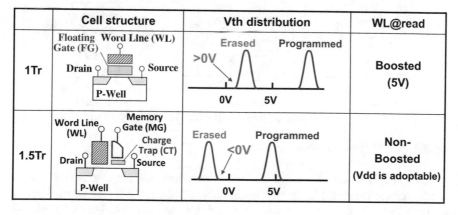

Fig. 7.3 Advantages of a 1.5Tr (split-gate) cell

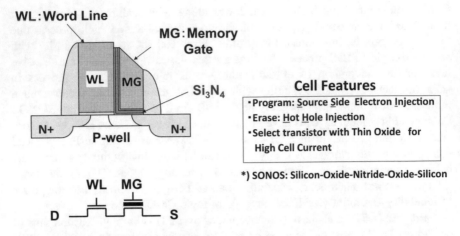

Fig. 7.4 The SONOS split-gate cell structure

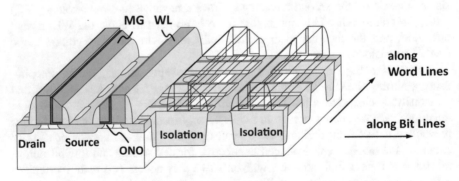

Fig. 7.5 Overview of the SONOS split-gate cell structure

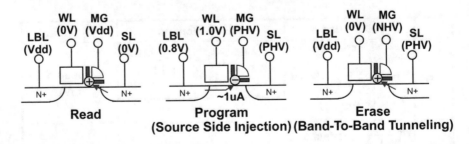

Fig. 7.6 Simplified cell operation

In Figs. 7.6, 7.7 and 7.8, simplified read-, program-, and erase-operation voltage conditions are described, respectively. In read operation, the fastest random-access read can be realized thanks to the low-voltage split-gate structure because word-line (WL) and bit-line voltages are the logic power supply (V_{dd}). For program operation,

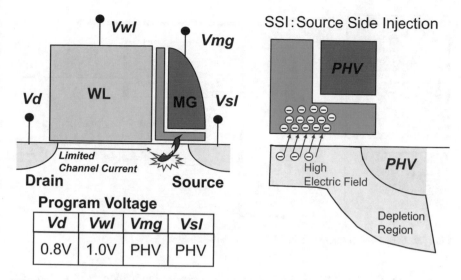

Fig. 7.7 Mechanism of programming

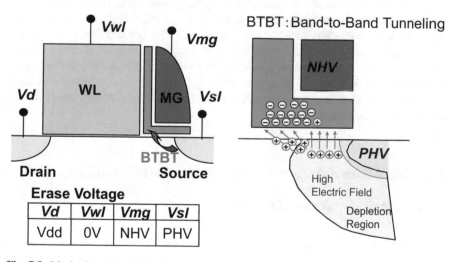

Fig. 7.8 Mechanism of erasing

SSI can realize fast and low-power programming with small cell current (approximately 1 μA). Finally, for erase operation, hot-hole injection by band-to-band tunneling (BTBT) is used. Especially, it contributes fast erase time for small erase unit.

The fabrication process of general-use flash memory, such as NAND flash memory, is mainly optimized for the shrinkage of memory cells and HV transistors dedicated for rewrite operations. In contrast, in the fabrication process of eFlash for a high-end microcontroller unit, it is important to balance sufficient performance of the logic CMOS transistors with memory-cell reliability.

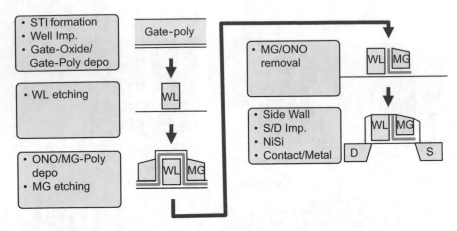

Fig. 7.9 SG-MONOS memory cell fabrication—process flow

An example of a SONOS split-gate eFlash-memory-fabrication process is shown in Fig. 7.9 [1]. At first, the gate oxide film used for the select transistor and peripheral logic CMOS are formed. The first poly-Si film, which is commonly used for select transistor and peripheral logic CMOS, is deposited. Then the oxide–nitride–oxide (ONO) film and the second poly-Si film are deposited to form the memory gate (MG), which is formed using anisotropic etching. A standard CMOS process is used for the remaining process step successively. A SONOS split-gate eFlash memory is fabricated first. Reversing the order may cause peripheral logic CMOS-performance degradation because logic CMOS suffers thermal stress during memory-cell fabrication.

7.2 Basic Memory Array Architecture and Operations

7.2.1 Basic Array Architecture

Figure 7.10 shows the voltage conditions in read operation. The cell at the cross-point of the selected WL and the selected BL is selected. Core-logic supply voltage, V_{dd}, is applied only to the selected WL and BL. Because there is no need to

Fig. 7.10 Read-voltage condition

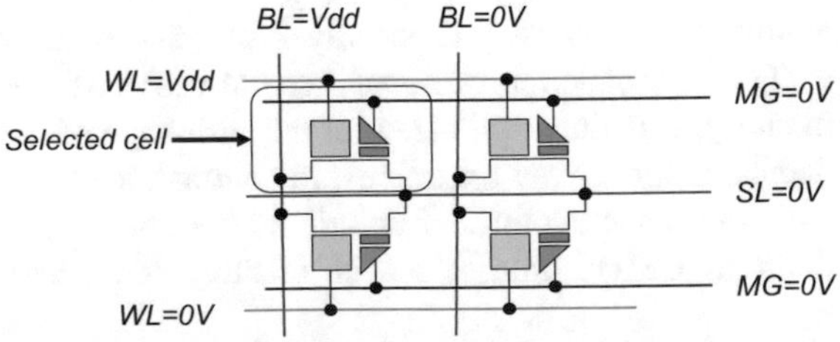

Fig. 7.11 Program-voltage condition

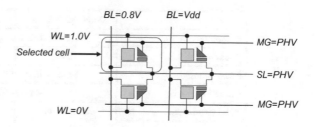

Fig. 7.12 Erase-voltage condition

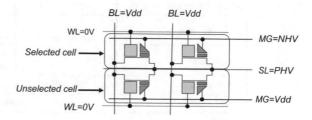

Fig. 7.13 Program-disturb modes

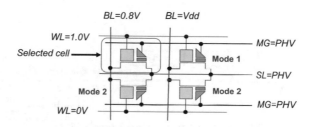

boost the WL in read operation, the fast-read path can be built with only fast low-voltage logic CMOS transistors. Furthermore, it is noteworthy that this array architecture is completely free from read disturb because MG voltage is 0 V throughout the read operation.

Figure 7.11 shows the voltage conditions during the program operation. High positive voltages (PHV [approximately 4–10 V]) are applied to the MG and the SL. In order to realize efficient SSI programing, low WL voltage (1 V) is applied to the selected WL.

Figure 7.12 shows the voltage conditions during the erase operation. To realize BTBT erase, negative high voltage (NHV [approximately −4 to −6 V]) and PHV are applied to the selected MG and SL, respectively.

In general, high voltages for program operation, which are also applied to unselected cells, cause the program disturb. Figure 7.13 shows disturb mode 1 and 2, which lead to an undesirable V_{th} shift at the unselected memory cells caused by MG and SL selection voltage.

Figure 7.14 shows the code and data-flash macro specifications for automotive applications. Current eFlash designs for MCU incorporate two types of eFlash macros: code flash and data flash. The code-flash macro is used to store the boot and application codes, which require large capacity and fast read operation at a wide

	Code-Flash Macro	Data-Flash Macro
Process	40nm CMOS, 2-polySi, 6-Cu, 1-Al	
Capacity	2MB	64KB
Power Supply	Core (VDD) 1.25V , I/O 2.7-5.5V	
Operating Temp.	-40~170 °C (Tj)	
Read Bandwidth	128bit + 10bit	32bit + 7bit
Random Read Freq.	160MHz	10MHz
Block Size / program unit	32KB / 256B	64B / 4B
Program Time	Rewrite (max) ~25s/2MB	350us/4B (average after 10M cyc.)
Erase Time		5ms/64B (average after 10M cyc.)
P/E Endurance	10K cycles	10M cycles
Data Retention	20years	

Fig. 7.14 Code and data-flash macro specifications for automotive applications

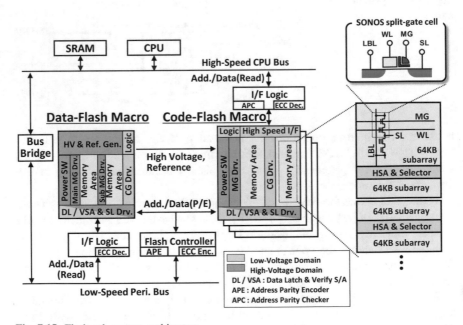

Fig. 7.15 Flash sub-system architecture

temperature range. The data-flash macro is the substitute for the external EEPROM chip in order to reduce system cost, and must meet the requirement for high program/erase endurance at wide temperature range. The fast erase speed feature of SONOS split-gate by BTBT enables small erase unit for data-flash macro with very short erase time, which is suitable to endure against frequent re-write.

Power Distributor	Control Unit	Power Distributor	Control Unit
LATCH & Verify-S.A.		LATCH & Verify-S.A.	
Extra memory area		Extra memory area	
Hierarchical Sense Amplifier		Hierarchical Sense Amplifier	
Extra memory area		Extra memory area	
64KB (= 32KB Erase Block x 2)		64KB	
Hierarchical Sense Amplifier		Hierarchical Sense Amplifier	
64KB		64KB	
64KB		64KB	
Hierarchical Sense Amplifier	Low Voltage Row Decoder / Driver	Hierarchical Sense Amplifier	High Voltage Row Decoder / Driver
64KB		64KB	
64KB		64KB	
Hierarchical Sense Amplifier		Hierarchical Sense Amplifier	
64KB		64KB	
64KB		64KB	
Hierarchical Sense Amplifier		Hierarchical Sense Amplifier	
64KB		64KB	
64KB		64KB	
Hierarchical Sense Amplifier		Hierarchical Sense Amplifier	
64KB		64KB	
64KB		64KB	
Hierarchical Sense Amplifier		Hierarchical Sense Amplifier	
64KB		64KB	
64KB		64KB	
Hierarchical Sense Amplifier		Hierarchical Sense Amplifier	
64KB		64KB	
64KB		64KB	
Hierarchical Sense Amplifier		Hierarchical Sense Amplifier	
64KB		64KB	
I / O		I / O	

Fig. 7.16 Code-flash architecture

An example of SONOS split-gate eFlash sub-system architecture is shown in Fig. 7.15. The number of code-flash macros is selectable according to the required code capacity. The data-flash macro has HV and reference-voltage generators for program and erase operations, which are shared by the data and code-flash macros to reduce the total area of eFlash macros in a chip. In the high voltage-circuit domain, high voltage is handled by I/O transistors, thus eliminating the need for a special HV device dedicated for Flash design. The flash controller interprets the incoming commands and controls the program and erase operations. For the safety function, data parity bits are implemented in both of the code-flash macros and the data-flash macro to support ECC (Error Correction Code) functionality. In addition, 1 address parity bit/128-bit read unit is introduced in the code-flash macros, which supports the address decoder-fault detection required by ISO26262.

Figure 7.16 shows an example of 2-MB code-flash macro configuration. For fast random-access read, the hierarchical sense amplifiers (HSAs) are placed in the memory array. A low-voltage row decoder and driver are placed at the center of the flash macro. Extra memory area is used as a data-storage area other than the user-memory area for boot code, redundancy, replacement information, product information, trimming data, and so on.

Fig. 7.17 A micrograph of 28-nm embedded SONOS split-gate eFlash macros

A typical SONOS split-gate eFlash macros are shown in Fig. 7.17 [2, 3]. It includes two code-flash macros and a data-flash macro. Multiple eFlash macros are embedded in a chip to implement multiple memory banks for MCU with multi-core CPU. OTA (over-the-air) applications also require multiple eFlash macros to store back-up data.

7.2.2 Read Operation

In the read-path design for a high-performance eFlash macro, isolating the high-speed read path from the HV program/erase path is preferable because

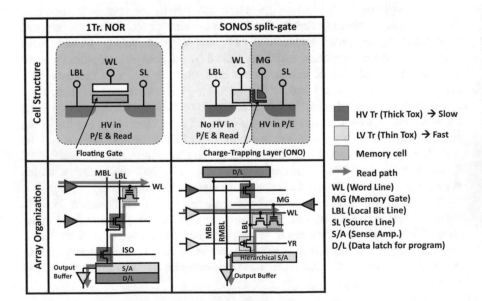

Fig. 7.18 SONOS split-gate cell structure and array organization

Fig. 7.19 Read-current trend
of SONOS split-gate eFlash

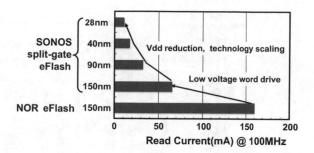

eliminating the HV transistors from the read path is effective to reduce delay time. Figure 7.18 shows the comparison of cell structure and array organization between SONOS split-gate type and 1Tr NOR type. In a 1Tr NOR-type cell array, HV transistors are generally involved in the read path because the boosted WL voltage is necessary to obtain sufficient memory-cell current for high-speed read while preventing an over-erase issue. However, the SONOS split-gate cell with a negative-erase threshold voltage enables a non-boosted simple read-path design realizing a separation between the low-voltage domain for read and the HV domain for program and erase operations.

Figure 7.19 shows the read-current consumption comparison between NOR eFlash and SONOS split-gate eFlash. NOR eFlash consumes more read current due to the boosted WL. A boosted WL causes larger current consumption because a charge-pump circuit for WL voltage consumes more operating current. In contrast, the read path of the SONOS split-gate eFlash does not need boosted voltage. Read-current consumption of the SONOS split-gate eFlash is reduced according to V_{dd} reduction due to CMOS-technology scaling.

Figure 7.20 shows an example of fast random-access read architecture. In the multi-divided memory array mats, the sensing and reading path is hierarchically organized to achieve a high-speed read operation. Therefore, the differential sense amplifiers with reference current source (I_{ref}) are placed in memory mats. Read-access time is divided into four phases as follows.

(1) $T1$: Address-decode time
(2) $T2$: WL activate and LBL (local bit line) pre-charge time
(3) $T3$: Sense amplifiers activate time
(4) $T4$: GBL (global bit line) bus drive time

Figure 7.21 shows the timing chart of fast read access. At the rising edge of the chip-clock, the read address is latched and decoded (T1). After address decoding, only one selected WL is activated, and the LBLs are pre-charged to V_{dd} (T2). After the sense amplifiers activate time (T3), they complete the sensing operation, and the read data are output to the GBL bus corresponding to the memory cell's V_{th} (T4).

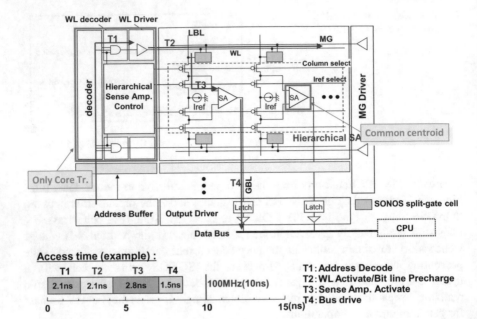

Fig. 7.20 Fast random-access read architecture

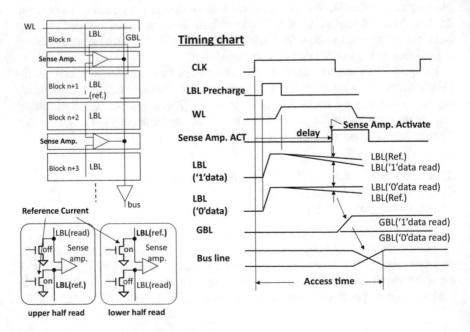

Fig. 7.21 Fast read access read timing chart

7.2.3   Erase/Programming Operations

For erase operation, the advantage of a split gate-type cell over 1Tr-type cell is that the negative threshold-voltage setting is available. A write-back operation to avoid an over-erase issue, which is essential in a 1Tr-type cell, is not required. Figure 7.22 shows a basic erase flow adopted for SONOS split-gate eFlash memory. Although an erase-verify operation is not mandatory for split-gate type cells, it is included in erase flow to prevent excessive erase-voltage stress.

Figure 7.23 shows erase-pulse waveforms using hot-hole injection by BTBT. Voltage stress is relaxed by incremental step pulse. Because the back-gate bias of the selected erase block is 0 V, which is identical to that of unselected erase block, there is no need to separate the back-gate (Pwell) between adjacent erase blocks. Combined with the fast-erase feature of BTBT operation, SONOS split-gate eFlash memory is suitable to realize the data-flash macro with a short erase time with a small erase unit, such as 32 B, thus well replacing the external EEPROM chip.

As for program operation, SONOS split-gate eFlash memory adopts efficient programming by source-side injection. The basic program operation flow is shown in Fig. 7.24.

Figure 7.25 shows voltage waveforms of SSI programming. An incremental step pulse is also adopted for program operation to relax voltage stress the same as during the erase operation. The average program current (I_{pgm}) is suppressed to ~1 μA/cell thanks to SSI programming.

Fig. 7.22 Basic erase flow of SONOS split-gate Flash memory

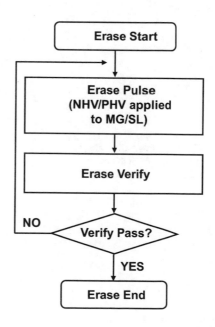

Fig. 7.23 Erase-voltage
waveforms of SONOS
split-gate flash memory

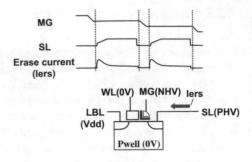

Fig. 7.24 Basic program
flow

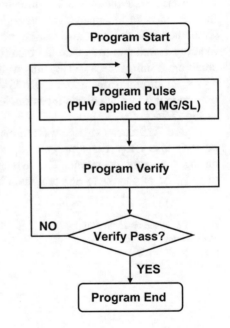

Fig. 7.25 Program-voltage
waveforms of SONOS
split-gate eFlash memory

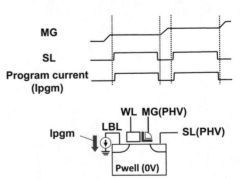

Good normal V_{th} distributions are observed for erased and programmed memory-cell arrays as shown in Fig. 7.26. The advantageous SONOS property against defects provides excellent V_{th} distributions.

In SG-MONOS flash memory macro, there are charge pumps (CPs), voltage regulators, and voltage switch circuits to apply appropriate high voltage to memory cells according to operation modes. Figure 7.27 shows a schematic diagram of HV supply paths of SONOS split-gate eFlash memory. There are two CPs to supply positive high voltages (PHVs) for program memory-gate (MG) voltage, program/erase source-line (SL) voltages, and a charge pump to supply negative high voltage (NHV) for erase-MG voltage.

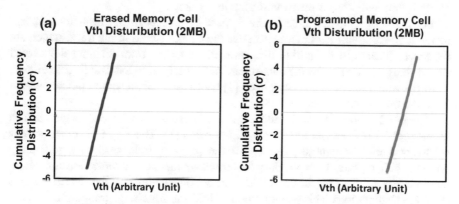

Fig. 7.26 **a** Erased memory-cell V_{th} distribution and **b** programmed memory-cell V_{th} distribution

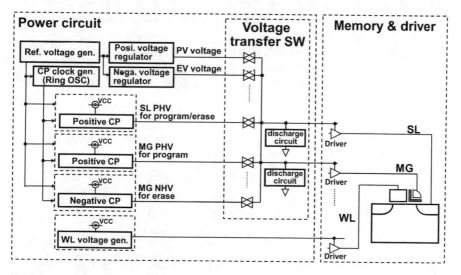

Fig. 7.27 A high-voltage supply path of SONOS split-gate eFlash memory

7.2.4 Reliability

Various defects in the insulator film significantly affect the reliability of flash memory. Especially for FG devices, even a single defect causes fatal storage-charge loss because a FG is made of conductive material. Not only hard defects, which are introduced in the fabrication process, but also soft defects occur by stress of repeated program/erase cycles, thus causing reliability to deteriorate. A representative phenomenon caused by a soft defect is the charge loss by SILC (stress-induced leakage current) arising from a hole trap in the insulator film. Although a flash MCU chip, including hard defects, can be repaired or rejected during testing, soft defects are difficult to find during testing before shipment because they will appear after many rewrite cycles.

For SONOS split-gate eFlash, the impact of defects in the insulator film are fewer compared with FG devices because the charge stored in the traps, which are separately located in the nitride film, prevents a defect from draining the overall stored charges. To estimate the data-retention property of SONOS split-gate eFlash, a simple estimation method is proposed [1] based on a data-retention model derived from thermionic emissions.

Figure 7.28 shows the data-retention characteristics of SONOS split-gate eFlash memory at 150 °C as measured by a TEG device [1]. The V_{th} of the erase state is relatively stable. In contrast, the V_{th} of the program state changes significantly within 1 h. This short-term V_{th} drop is caused by the electron/hole recombination, the recovery of the interface state (silicon/bottom-oxide) and the charge de-trap from the shallow trap. This phenomenon is intrinsic for charge-trapping memories [4, 5]. After that, the V_{th} gradually decreases mainly by thermionic emission shown in Fig. 7.29 [1].

For automotive use, long-term data retention, such as 20 years, is required. However, long-term evaluation and a great number of samples are needed to prove the high-reliability characteristics.

Particularly, it was very difficult to predict the data-retention lifetime although it is the most significant factor of eFlash reliability. A data-retention model based on the thermionic emissions is proposed, and a simple estimation method for data

Fig. 7.28 Data retention by SONOS split-gate memory (150 °C)

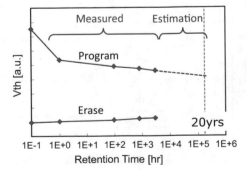

Fig. 7.29 Data-retention
mechanism for the long term
(>1 h)

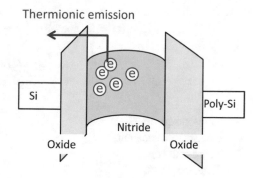

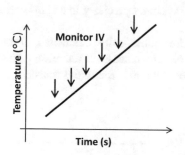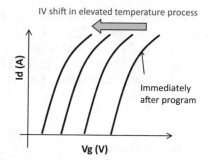

Fig. 7.30 *I–V* curve measurement in high-temperature profile

retention over a long period and a wide temperature range has been developed for
SG-MONOS flash memory [1]. To estimate the long-term data-retention characteristics dominated by thermionic emission, trap level and the density-distribution
profile in nitride film are required. They are derived by following two steps.

(1) SONOS split-gate cell V_{th} shift is measured in high-temperature profile. An
    example of measurement condition is shown in Fig. 7.30. The measured
    temperature range is 200–600 °C and is raised at a constant rate. *I–V* curves are
    acquired at fixed intervals, and the V_{th} shift at each temperature are obtained.
(2) The V_{th}-shift dependency on temperature is converted to the relationship
    between trap energy and trap density. Temperature can be converted to trap
    energy using Eqs. (7.1)–(7.3). The V_{th} shift is converted to trap density by
    using the differential value derived from Eq. (7.4).

$$\Phi = kT(-2 + W(x)) \tag{7.1}$$

$$x = \frac{AT^2 \exp(2)}{B} \tag{7.2}$$

$$W(t) = \ln(t - 4) - \left(1 - \frac{1}{\ln(t)}\right) \cdot \ln(\ln(t)) \tag{7.3}$$

$$\frac{dV_{th}(\Phi)}{d\Phi} \tag{7.4}$$

Φ Trap-energy level,
k Boltzmann constant,
T Temperature,
A Time constant,
B Rate of temperature increase

Trap-distribution profile in nitride can be measured precisely in a short period of measurement as shown in Fig. 7.31.

The data-retention characteristics can be numerically calculated from the trap-density profile using Eqs. (7.5) and (7.6). The calculated retention characteristics is in good agreement with the measured data after the initial transient 1 h as shown in Fig. 7.32.

Fig. 7.31 Trap-distribution profile in nitride

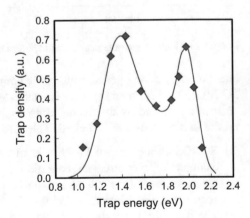

Fig. 7.32 Data-retention property: calculated versus measured

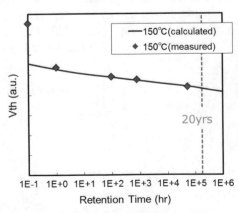

$$\frac{dZ(x, \Phi_t)}{dt} = -\frac{Z(x, \Phi_t)}{\tau_{TH}(x, \Phi_t)} \qquad (7.5)$$

$$\tau_{TH}(x, \Phi_t) = \tau_{TH0} \exp\left(-\frac{q\Phi_t}{kT}\right) \qquad (7.6)$$

$Z(x, \Phi_t)$ Trap-density profile in nitride,
x Trap position,
Φ_t Trap level in nitride
τ_{TH0} Time constant of thermionic emission

7.3 Advanced Design Techniques to Expand Applications

In order to expand the application range of embedded flash memory, high-performance features—such as fast read, program, and erase operations are required. Advanced design techniques to achieve high-speed read and fast erase/program operations are described in the following Sects. 7.3.1 and 7.3.2, respectively.

For flash-memory design, HV generation for erase/program operations is always an unavoidable challenge. In Sect. 7.3.3, the significant limitations for eFlash, e.g., charge-pump design, HV device restriction, and the noise issue, are discussed and the measures for them described.

7.3.1 Read Techniques

As mentioned previously, SONOS split-gate cells have intrinsically fast random-access read characteristics. However, as CPU frequency increases, eFlash random-access read speed requirements continually increase. In this section, two techniques—offset-cancelling sense amplifier and WL-overdrive—to further improve the random-access read speed of SONOS split-gate eFlash are introduced.

Offset-cancelling sense amplifier
Figure 7.33 shows a conventional latch-type voltage sense-amplifier circuit. In general, a sense amplifier has the offset caused by the pair transistors' mismatch. The offset causes a longer sense time because if there is the offset during sense time, the sense amplifier (SA) requires additional offset voltage to amplify. Therefore, the sense time becomes longer than without offset.

In order to shorten the sense time, SA with digital-offset cancellation (SA-DOC) is adopted [6, 7]. Figure 7.34 shows SA-DOC circuits. The offset-cancel current

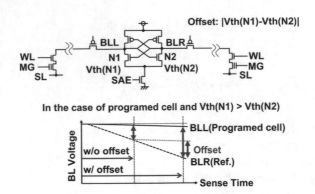

Fig. 7.33 Conventional-voltage sense amplifier

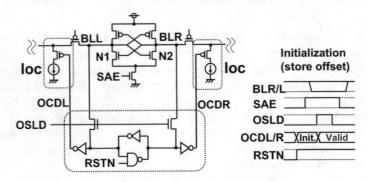

Fig. 7.34 Sense amplifier with digital-offset cancellation (SA-DOC)

(I_{oc}) is set to compensate for half of the maximum offset. During the initialization of the eFlash system, e.g., power-on sequence, the sense amplifier senses its own offset, and the result is stored in the dedicated latch digitally.

Figure 7.35 shows the improvement of sense time by SA-DOC. By using SA-DOC, the sense time can be shortened because the I_{oc} current compensates for the offset.

Figure 7.36 shows a measured random-access shmoo plot for the 40 nm—generation code-flash macro at $T_j = 170\ °C$. Random read access frequency shows around 135 MHz w/o SA-DOC and over 160 MHz random access is achieved by using SA-DOC. Thus, frequency improved by 26%, in other words, sense time decreased 1.5 ns. Furthermore, by reading from two code macros (128bit × 2), and the maximum random access throughput reaches 5.1 GB/s.

Word-line overdrive

In eFlash-circuit design, to achieve both sufficient memory cell current for high-speed read and high reliability of the peripheral circuit is a significant challenge especially for high-end flash MCU for automotive applications.

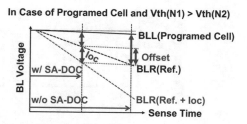

In Case of Programed Cell and Vth(N1) > Vth(N2)

Fig. 7.35 Improvement of sense time

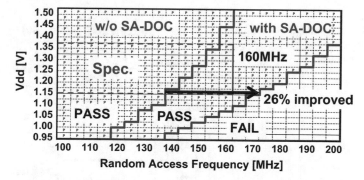

Fig. 7.36 Measured random access

In conventional flash memory, the WL voltage of selected memory cells are supplied from the power-supply voltage for the logic circuit. Namely, V_{DD} is supplied regardless of temperature as the dashed line shown in Fig. 7.37a. Read memory-cell current (I_{cell}) is decreased in low temperature because the threshold voltage of the memory cell increases at low temperature as the dashed line shown in Fig. 7.37b.

WL overdrive is one of the candidates to increase cell current. However, simple WL overdrive degrades the reliability of the peripheral circuits, especially at high temperature, due to the temperature dependence of TDDB lifetime.

To achieve both sufficient I_{cell} for high-speed read and reliability of the peripheral circuit throughout a wide temperature range, e.g., −40 to 170 °C, a WL overdrive with negative temperature dependence is introduced [2, 3]. As the solid line shown in Fig. 7.37a, the negative temperature dependence on WL voltage (VWL) is applied. It increases the I_{cell} at low temperature so that enough I_{cell} for high-speed read is obtained throughout wide temperature range as the solid line shown in Fig. 7.37b.

Figure 7.38 shows the schematic diagram of a WL-voltage generator and the voltage-transferring path to the memory cell. The reference voltage, V_{ref}, with negative temperature dependence is input to the voltage down converter (VDC). VDC generates WL voltage with negative temperature dependence, and it is applied to selected memory cells through a WL driver.

Fig. 7.37 The WL overdrive with negative temperature dependence

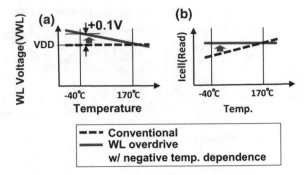

Fig. 7.38 WL overdrive voltage generator and the voltage-supply path

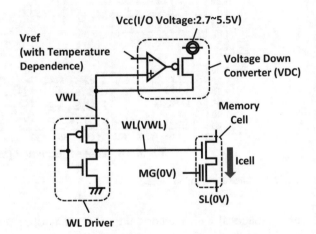

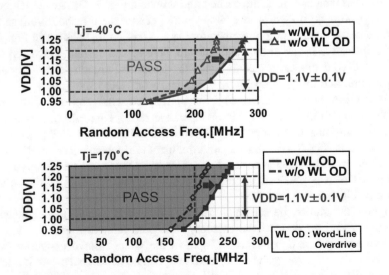

Fig. 7.39 Measured random access

Thanks to the technique of SA-DOC with WL overdrive having negative temperature dependence, 200-MHz random-access read is achieved as shown in Fig. 7.39 [2, 3].

7.3.2 Program/Erase Techniques

In this section, two techniques for SONOS split-gate eFlash to improve program and erase speed are introduced.

APCCS (Adaptable program current-control scheme)
Figure 7.40 shows a technical issue of program operation in 40-nm process. We observed minority slow cells in the relationship between the number of program pulses and the number of programmed units per step. The program time is limited by minority slow cells when we used a constant program current for all cells and every program step.

In order to improve the program speed, an APCCS is proposed [6, 7]. Figure 7.41 shows the block diagram of the APCCS. One of the two latches stores program data from the flash controller, and the other one stores a verification result from verify SA. The fail-bit counter (FBC) stores the fail-bit count after verification from all data latches. These latches and the output from the FBC combine the three types of program current, I_0, I_1, and I_2, to redistribute the program current. Figure 7.42 shows the progress of APCCS from the initial erase state to the programmed state. At first, the program current is set to $I_0 + I_1$ for erased cells. After several program pulses are applied and the FBC becomes less than a certain threshold value (NF) at the verification, the program current for the failed cells increases to $I_0 + I_1 + I_2$ to increase the program speed. In contrast, for the pass cells, the program current is set to I_0 to keep the pass state to secure successful completion. NF is set to the number of program unit cells divided by 4, for example. This is optimized considering the total program current and charge-pumping current.

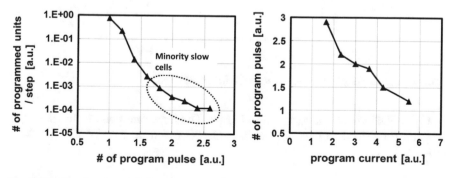

Fig. 7.40 Minority slow cells in program operation

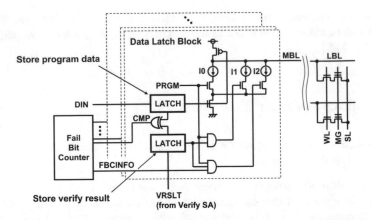

Fig. 7.41 Block diagram of adaptable program cell current-control scheme (APCCS)

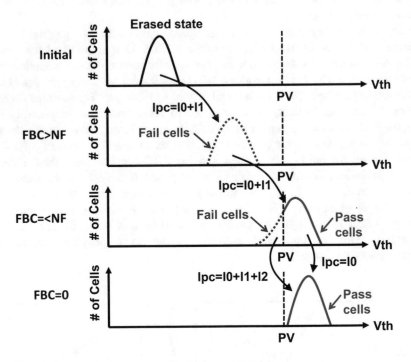

Fig. 7.42 Adaptable program current-control scheme (APCCS)

Figure 7.43 shows the experimental result. The important point is that the worst program time decreases by 50% at $T_j = 170\ ^\circ\text{C}$ owing to the APCCS in 40-nm process.

Fig. 7.43 Improvement of
program time

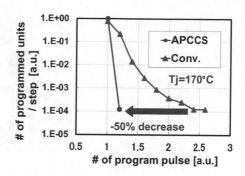

IES (intelligent erase scheme)

An advanced design technique for the erase operation is described. In the conventional erase scheme shown in Fig. 7.44, the width of the constant-erase pulse is adjusted to the worst condition.

In order to shorten the erase time, an intelligent erase scheme (IES), shown in Fig. 7.45, is adopted [6, 7]. VCP voltage is monitored by the voltage detector to observe the change of erase current (I_{ers}) because VCP voltage strongly depends on I_{ers} consumption. The sequence controller adjusts the erase-pulse width using the output of the voltage detector.

Figure 7.46 shows the operation waveforms of IES. At first, the VCP-voltage decreases due to large I_{ers}-current consumption. After that, when I_{ers} decreases enough and VCP voltage recovers, the sequence controller finalizes the current step and starts the next step. In other words, we can optimize the pulse width dynamically by every step.

Figure 7.47 shows the experimental result. A dashed line shows the conventional and a solid line shows the IES. The IES significantly decreases erase-pulse time at $T_j = 170$ °C by 60%.

Figure 7.48 shows the program- and erase-endurance characteristics of 40-nm SONOS split-gate data-flash macro at $T_j = 170$ °C. The APCCS and IES enable shorter program and erase time, respectively. It means that the electrical stresses on memory cells can be relaxed. Thanks to the intrinsically high reliability of SONOS

Fig. 7.44 Conventional erase
scheme

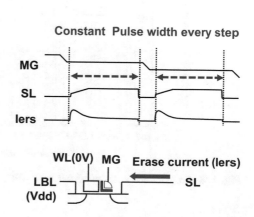

Fig. 7.45 Intelligent erase
scheme (IES)

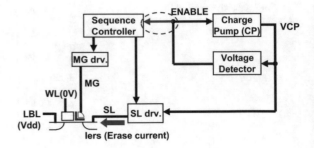

Fig. 7.46 Waveforms of IES

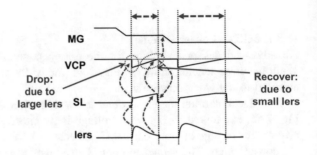

Fig. 7.47 Improvement of
erase time by IES

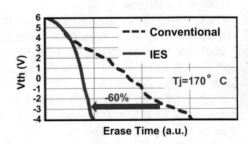

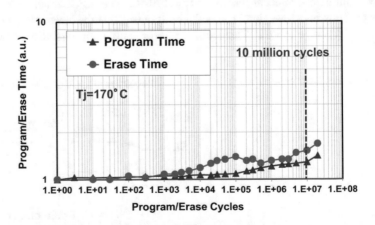

Fig. 7.48 Program and erase endurance with APCCS and IES

split-gate cells and the electrical-stress relaxation by APCCS and IES, the data-flash macro can achieve the endurance characteristics >10 million (10 M) cycles at $T_j = 170\ ^\circ\mathrm{C}$.

7.3.3   Charge-Pumping Circuitry

Figure 7.49 shows a schematic diagram of a conventional charge-pumping circuit. It is constructed in multiple stages and has a HV detector circuit. Each stage includes a diode device, a pump capacitor, and a capacitor driver. Odd stages and even stages are driven by complementary clocks. There are two significant challenges for HV circuit design for eFlash as follows.

(1) From the viewpoint of total process cost, any HV circuits, including the CPs in eFlash macros, should be composed of the standard HV transistors and MOS capacitors used for existing I/O circuits and analog circuits in a flash-MCU with a power supply of 5 or 3.3 V. Accordingly, stress voltage and stress time applied to HV transistors and MOS capacitors must not exceed the reliability criteria of these existing devices, thus posing severe limitation especially for charge-pump design.

(2) Charge-pump operating noise should be suppressed to avoid electro-magnetic interference (EMI) issue. Charge-pump circuits consume large amounts of peak current for charging and discharging the pump capacitors. Rapid changes in the consumption of current occur, and EMI noise is emitted. This may interfere with wireless communications and cause malfunction.

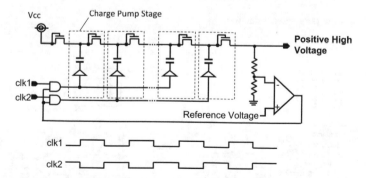

Fig. 7.49 A conventional charge-pump circuit

In general, the area size of the CPs is dominant in the HV and reference-voltage generator blocks. Figure 7.50 shows the circuit diagram of a conventional CP. A single HV MOS capacitor is used in each of the low-voltage stages, and two serially connected HV MOS capacitors are used in each of the HV stages in order to mitigate the biased high voltage on the capacitors. However, the serially connected HV MOS capacitors cause quite a large area penalty. To solve this problem, lateral metal-oxide-metal (MOM) capacitors and poly-insulator-poly (PIP) capacitors stacked over the HV MOS capacitors are adopted [6, 7]. Figure 7.51a, b show the capacitor structure for the low- and high-voltage stages. A PIP capacitor, MOM capacitors with narrowly spaced metals, and a HV MOS capacitor are connected in parallel for low-voltage stages, while the HV stage is composed of widely spaced MOM capacitors in parallel with serially connected PIP and MOS capacitors.

Figure 7.52 compares the normalized charge pump area by conventional and proposed configurations. The charge-pump area is reduced by 44% using MOM

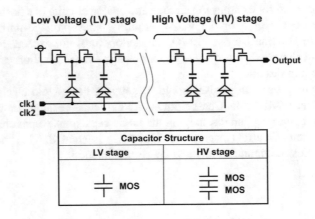

Fig. 7.50 Charge pump-area reduction

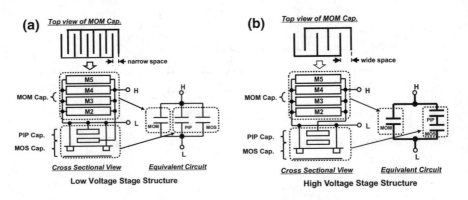

Fig. 7.51 **a** Low voltage stage structure **b** High-voltage stage structure

	CP Area (normalized)	Capacitor Structure	
		LV stage	HV stage
Conventional	1	⊥ MOS	MOS / MOS
Proposed	0.56	MOM (narrow) / PIP MOS	MOM (wide) / PIP MOS

Fig. 7.52 Improvement of the charge-pump area

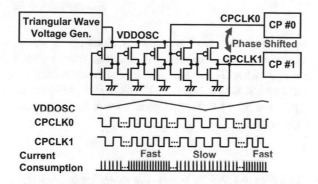

Fig. 7.53 Phase shift and SSCG for charge-pump operating clock-generation

and PIP stacked over the HV MOS capacitors while keeping a sufficient margin to the breakdown voltage.

Next, circuit techniques to reduce electronic-magnetic interference (EMI) are described. Low EMI noise operation is increasingly required for Flash MCU chips because of new applications such as in-field program updates by OTA by wireless communications. Varying charge pump-current consumption emits RF noise, thus interfering with incoming RF signals.

One of the well-known circuit technique to disperse the current-consumption peak is the pump clock-phase shift as shown in Fig. 7.53. Phase-shifted clocks, CPCLK0 and CPCLK1, are supplied to charge pump0 (CP #0) and charge pump1 (CP #1), respectively. Furthermore, spread-spectrum clock generation (SSCG) is effective to reduce peak power in the frequency domain [2, 3]. As shown in Fig. 7.53, the frequency of the clock signals is intentionally spread by supplying a triangular wave voltage at a power supply of the ring-oscillator circuit, VDDOSC. Because the frequency of current consumption-peaks changes follows the variation of VDDOSC, the peak power in the frequency domain is suppressed.

Figure 7.54 shows an FFT analysis of simulated current-consumption waveforms in a single macro-operation case and dual macro-operation case. The FFT

Fig. 7.54 FFT waveforms of
the charge pump-consumption
current

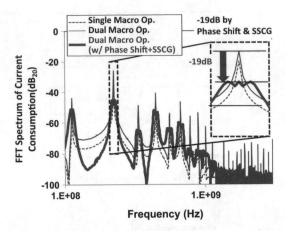

spectrum of the dual-macro operation without clock-phase shift and SSCG for
charge pump operating clock-generation is simply two times compared with that of
a single-macro operation. The peak value of the dual-macro operation with
clock-phase shift and SSCG is reduced by 19 dB, which is even smaller than that of
single macro operation by −13 dB. The peak powers in the frequency domain are
successfully suppressed.

7.4   Scalability of SONOS Split-Gate eFlash Memory

Although many types of eFlash-memory cells have been proposed and manufactured, the number of those actually adopted in commercial products tends to
drastically decrease along with technology scaling beyond 90 nm (Fig. 7.55).
Especially in MCU applications, they have been converging into four types as

Fig. 7.55   eFlash-technology
selection along with process
scaling

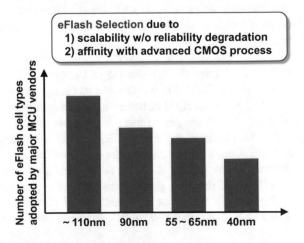

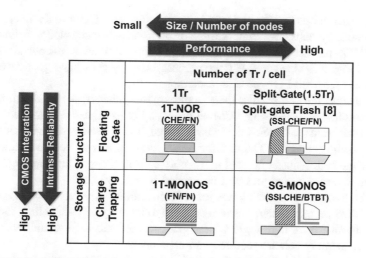

Fig. 7.56 Typical eFlash cells in 90 nm and beyond in MCU applications [8]

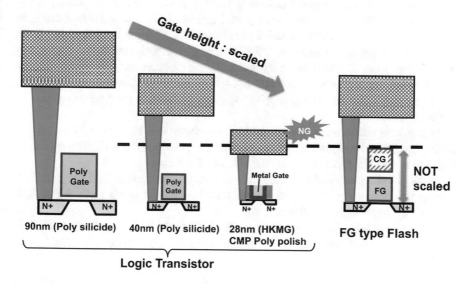

Fig. 7.57 Gate-height trend in logic transistors

illustrated in Fig. 7.56. The main factors of scalability, or "survivability" beyond 90 nm are (1) high scalability without degrading reliability, and (2) high affinity with advanced-logic CMOS process where eFlash-memory cells are embedded.

The gate height of logic CMOS transistors have been getting lower along with scaling, which makes it difficult to integrate certain types of eFlash into advanced-logic CMOS process due to its high cell profile, as shown in Fig. 7.57. In contrast, the low-profile eFlash cell shows high affinity with advanced-logic CMOS process. This means that the cell profile can be a decisive factor in the integration of eFlash-memory cells into logic CMOS process in advanced node. Affinity with logic CMOS process has become much more important than ever especially beyond 40 nm. From this viewpoint, CT-type eFlash-memory cells have great advantages over compared with FG-type eFlash-memory cells.

Aside from the cell profile, there could emerge other issues in eFlash-memory integration into advanced-logic CMOS process. For example because the thickness of dielectric films between interconnect metal layers become thinner along with process scaling, and the TDDB lifetime of these films can become critical considering the high voltages applied to memory cells in program and erase operations. Some new design techniques will be needed beyond 40 nm.

In the scalability consideration of eFlash, affinity with advanced high-speed logic CMOS process is significant in addition to memory-cell scalability. Especially for automotive uses, high logic CMOS performance and high reliability and scalability of memory cell are required simultaneously.

In 28 nm, a high-k metal gate (HKMG) process is widely adopted. In HKMG process, flash-memory cells are polished to the same height as logic CMOS transistors by chemical mechanical polishing (CMP). It is a very severe situation for floating-type cells because the memory cell height of FG-type cells is more than twice higher than that of logic CMOS transistors in the 28-nm process owing to its stacked gate structure. In contrast, SONOS split-gate eFlash memory-cell height is lower than that of FG-type cells because (1) the charge-trap layer for an SG-MONOS cell is thin SiN film; and (2) split-gates, word line (WL) and MG, are placed side-by-side. Therefore, SONOS split-gate eFlash is very advantageous in terms of scalability.

Figure 7.58 shows a comparison of initial memory-cell V_{th} distribution between a CMP planarized sample and an unpolished one. Good normal V_{th} distribution is kept even after CMP planarization for HKMG formation.

SONOS split-gate eFlash development is in progress toward 16-nm Fin-FET process [9]. Figure 7.59 shows an overview and a cross-sectional view of Fin-FET SONOS split-gate eFlash cell array.

Similar to a logic MOS transistor, a Fin-FET SONOS split-gate memory cell enjoys steeper I–V curve than a planar type cell as shown in Fig. 7.60a.

Fig. 7.58 Initial memory-cell V_{th} comparison between polished and unpolished samples

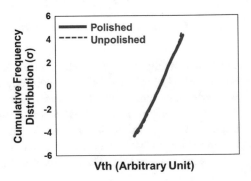

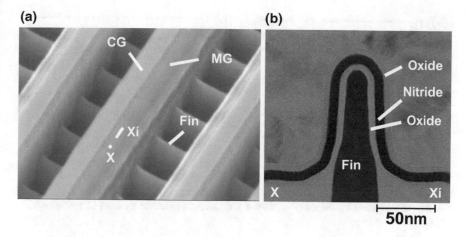

Fig. 7.59 A 16-nm Fin-FET SONOS split-gate eFlash: **a** overview and **b** cross-sectional view

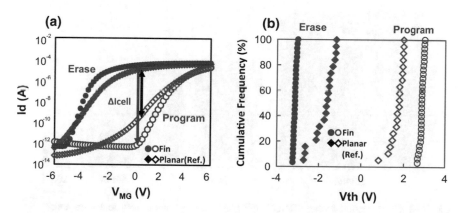

Fig. 7.60 The 16-nm Fin-FET SONOS split-gate eFlash characteristics. **a** I–V characteristics of erase/program memory cells. **b** Cumulative erase/program memory-cell V_{th} distributions

The Fin-FET SONOS split-gate eFlash can achieve faster read access thanks to wider memory cell-current window compared with planer type cell. Figure 7.60b shows the comparison between the width of memory-cell V_{th} distributions of Fin-FET/planer SONOS split-gate memory cells. The Fin-FET memory cell brings narrower V_{th} distribution compared with a planer one thanks to its fully depleted characteristics. Figure 7.61a shows the endurance characteristics. No remarkable degradation of program/erase (P/E) time is observed until 250 k P/E cycles. The data-retention characteristics are shown in Fig. 7.61b. The extrapolated characteristics predicts sufficient V_{th} window after 10-year data retention.

Consequently, SONOS split-gate eFlash matches well with advanced logic-CMOS fabrication techniques at 16-nm node, HKMG, and Fin-FET.

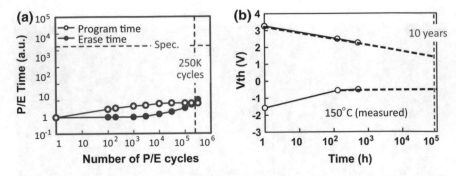

Fig. 7.61 Endurance and data retention characteristics of 16 nm Fin-FET SONOS split-gate eFlash: **a** memory cell V_{th} during P/E cycles. **b** Data retention characteristics at 150 °C after 250 k P/E cycles

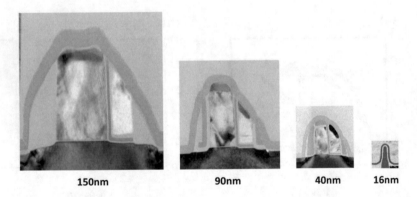

Fig. 7.62 Cross-sectional view SONOS split-gate memory cells at 150- to 16-nm nodes

Figure 7.62 shows a cross-sectional view of a SONOS split-gate eFlash-memory cell from the 150- to 16-nm technology nodes. Based on the proper operations of a Fin-FET SONOS split-gate memory cell above, it proves to be one of the important candidates for eFlash at 16- and 14-nm nodes and beyond.

7.5 Conclusion

A SONOS split-gate eFlash-memory cell, which is a split-gate memory cell with charge-trapping storage, presents the following attractive features: fast and low-power programming capability by SSI, fast and low-power read operation by boost-less WLs, and robust data reliability and scalability by charge-trapping structure. Therefore, SONOS split-gate structure is especially suitable for automotive applications because high-speed random-access read characteristics, a very low defect rate, and a harsh high-temperature durability are required. In addition, thin-nitride film storage realizes a low-profile cell and enables high compatibility with base logic CMOS for beyond 40 nm. The following advanced-circuit techniques were adopted to enhance the intrinsic advantages of SG-MONOS cells with reduction of the peripheral circuit area. The technique of employing a sense amplifier with digital offset-cancellation (SA-DOC)—combined with the technique of using a WL overdrive with negative temperature-dependence—yields a 200-MHz random read-access achievement in 28-nm SONOS split-gate eFlash macro. An APCCS can improve the maximum program time by controlling the program current of fail bits according to the number of fail bits in a program unit. The intelligent-erase scheme (IES) dynamically controls the erase pulse width by monitoring the voltage, which reflects the transition of the erase current, and enables a drastic decrease of the erase time. APCCS and IES contribute to not only improving the program/erase time but also to alleviating the electrical stresses on memory cells. Thanks to the intrinsically high reliability of SONOS split-gate cells and the electrical stress relaxation by APCCS and IES, the 40-nm data-flash macro can achieve the endurance characteristics of >10 million (10 M) cycles at $T_j = 170$ °C. An SSCG and a clock phase-shift technique solve the EMI noise issue caused by charge-pump operation during in-field programming. Fin-FET SONOS split-gate eFlash development is in progress toward 16-nm Fin-FET process. The Fin-FET SONOS split-gate memory cell enjoys a steeper I–V curve than the planar-type cell. Consequently, SONOS split-gate eFlash matches well with advanced logic-CMOS fabrication techniques at 16-nm node, HKMG, and Fin-FET, and it proves to be one of the important candidates for eFlash at 16- and 14-nm nodes and beyond. These achievements prove that SONOS split-gate eFlash-memory cell, as well as the circuit techniques described in this chapter, have sufficient potential to meet the tough requirements of future automotive applications.

References

1. Y. Kawashima, T. Hashimoto, I. Yamakawa, Investigation of the data retention mechanism and modeling for the high reliability embedded split-gate MONOS flash memory. *IEEE International Reliability Physics Symposium*, pp. MY.6.1–MY.6.5 (2015)
2. Y. Taito, M. Nakano, H. Okimoto, D. Okada, T. Ito, T. Kono, K. Noguchi, H. Hidaka, T. Yamauchi, A 28 nm embedded SG-MONOS flash macro for automotive achieving 200 MHz read operation and 2.0 MB/s write throughput at Tj of 170°C. ISSCC digest of technical papers, pp. 132–133 (2015)
3. Y. Taito, T. Kono, M. Nakano, T. Saito, T. Ito, K. Noguchi, H. Hidaka, T. Yamauchi, A 28 nm embedded split-gate MONOS (SG-MONOS) flash macro for automotive achieving 6.4 GB/s

read throughput by 200 MHz no-wait read operation and 2.0 MB/s write throughput at Tj of 170°C. IEEE J. Solid-State Circuits **51**(1), 213–221 (2016)

4. M. Janai, B. Eitan, A. Shappir, E. Lusky, I. Bloom, G. Cohen, Data retention reliability model of NROM nonvolatile memory products. IEEE Trans. Device Mater. Reliab. **4**, 404–415 (2004)

5. P.B. Kumar, E. Murakami, S. Kamohara, S. Mahapatra, Endurance and retention characteristics of SONOS EEPROMs operated using BTBT induced hot hole erase. in *Proceedings IRPS*, pp. 699–700 (2006)

6. T. Kono, T. Ito, T. Tsuruda, T. Nishiyama, T. Nagasawa, T. Ogawa, Y. Kawashima, H. Hidaka, T. Yamauchi, 40 nm embedded SG-MONOS flash macros for automotive with 160 MHz random access for code and endurance over 10 M cycles for data. ISSCC digest of technical papers, pp. 212–214 (2013)

7. T. Kono, T. Ito, T. Tsuruda, T. Nishiyama, T. Nagasawa, T. Ogawa, Y. Kawashima, H. Hidaka, T. Yamauchi, 40-nm embedded split-gate MONOS (SG-MONOS) flash macros for automotive with 160-MHz random access for code and endurance over 10 M cycles for data at the junction temperature of 170°C. IEEE J. Solid-State Circuits **49**(1), 154–166 (2014)

8. L.Q. Luo, Y.T. Chow, X.S. Cai, F. Zhang, Z.Q. Teo, D.X. Wang, K.Y. Lim, B.B. Zhou, J.Q. Liu, A. Yeo, T.L. Chang, Y.J. Kong, C.W. Yap, S. Lup, R. Long, J.B. Tan, D. Shum, N. Do, J. H. Kim, P. Ghazavi, V. Tiwari, Functionality demonstration of a high-density 1.1 V self-aligned split-gate NVM cell embedded into LP 40 nm CMOS for automotive and smart card applications. IEEE International Memory Workshop, pp. 165–168 (2015)

9. S. Tsuda, Y. Kawashima, K. Sonoda, A. Yoshitomi, T. Mihara, S. Narumi, M. Inoue, S. Muranaka, T. Maruyama, T. Yamashita, Y. Yamaguchi, D. Hisamoto, First demonstration of FinFET split-gate MONOS for high-speed and highly-reliable embedded flash in 16/14 nm-node and beyond. IEDM digest of technical papers, pp. 11.1.1–11.1.4 (2016)

Index

A

Adaptable Program Current Control
Scheme (APCCS), 231–234, 243
Adaptable Slope Pulse Control (ASPC),
201–204, 207
Additional mask, 181, 182, 184, 194, 196, 206
Add-on flash, 22
Advanced Driver Assistance System (ADAS),
2, 3
Affinity with logic CMOS process, 42
Array architecture, 153
Array sector, 155
Automotive, 7, 12, 13, 16–19, 21, 25, 26,
75–77, 80, 81, 111, 113, 114, 117, 119,
122, 127, 128, 148, 179, 180, 182, 186,
201, 204, 207

B

Band-to-Band Tunneling (BTBT), 213, 215,
221
Band-to-Band Tunneling (BTBT) hot hole
injection, 44–46
BCD, 148
Bluetooth®, 1
Breakdown voltage, 55, 57, 70
Byte (B), 22

C

Chanel Hot Electron (CHE) injection, 35
Channel FN electron ejection, 35, 40, 43
Channel FN hole injection, 44
Charge mismatch, 45, 46
Charge Pump (CP), 54, 55, 57, 84, 87, 91,
94–97, 101, 102, 104, 110, 113, 114,
116, 125, 223, 235–238
Charge-trapping, 179, 180, 211, 242, 243
Charge-Trapping(CT) structures, 31
Code flash, 8, 9, 12, 147
Code macro, 215, 217, 228

Code memory, 112, 113, 122, 127
Current sense, 190

D

Data flash, 12, 22, 26, 147
Data macro, 217
Data memory, 112, 113, 118, 120, 122, 127
Data retention, 148
2.5D/3D integration, 2
Defect, 35, 39, 42, 44–46, 48
Deplete, 179, 188, 192
DFM/DFT, 61, 62, 70
Digital-Signal Processor (DSP), 15
Disturb, 88, 89, 91, 113, 122, 123, 125, 126,
165
Drain/Source (D/S), 141
DRAM, 3

E

eFlash innovation, 3
eFlash macro, 13
eFlash MCU, 81
eFlash selection, 238
eFlash system, 29, 52, 62, 65–70
Electrically Erasable Programmable Read Only
Memory (EEPROM), 7, 12, 20, 31, 49,
75, 82, 83, 111–113, 118, 120, 121, 131
Embedded flash, 147
Embedded flash memory (eFlash), 3–5, 91,
111, 180–182, 184, 204–207
Embeddedness, 3, 8
Embedded SuperFlash Gen 1 (ESF1), 133
Embedded SuperFlash Gen 2 (ESF2), 133
Embedded SuperFlash Gen 3 (ESF3), 133
Embedded system, 1–3
Endurance, 148, 180, 203, 204, 215, 233, 235,
241–243
Endurance (program/erase cycle), 29, 39
Erase unit/block, 36, 37, 52, 53

© Springer International Publishing AG 2018
H. Hidaka (ed.), *Embedded Flash Memory for Embedded Systems: Technology,
Design for Sub-systems, and Innovations*, Integrated Circuits and Systems,
DOI 10.1007/978-3-319-55306-1

Error Correction Code (ECC), 62, 66–68, 151

F
FDSOI, 148
Field-Programmable Gate Array (FPGA), 15
Flash-MCU, 3, 209, 215, 224, 228, 235, 237
Floating Gate (FG), 179, 181, 183, 206
Floating Gate (FG) structures, 31
FN electron injection, 35, 36, 39–41, 43, 44,
        46, 48, 49, 51
FN tunneling, 182, 187, 193, 194

H
Hierarchical architecture, 217
High performance, 209
High reliability, 209, 228, 233, 240, 243
HV circuit design, 52, 54, 55, 57, 70
HV decoder, 160
HV management, 76, 99
HV transistor, 77–79

I
Idling P/E Management Unit (IPEMU), 205,
        206
Info rows, 156
Innovation, 7, 10, 11, 24–27
Integrated Device Manufacturer (IDM), 133
Intelligent Erase Scheme (IES), 233, 234, 243
Intelligent Slope Pulse Control Circuit
        (ISPCC), 201, 202
Internet of Everything (IoE), 1–3
Internet of Things (IoT), 2, 3, 147
IP, 17, 22

L
Level shift circuit, 56
Lightly Doped Drain (LDD), 77
Low cost, 179, 180, 184, 197, 206
Low power, 180–182, 203, 204
Low power design, 62

M
Macro, 21, 22
Macrocell, 76, 79, 90, 91, 98–100, 102, 106,
        111, 122
Magnetic-RAM, 3
Mask ROM, 9, 10
MCU, 180, 181, 204, 206
Memory array, 32, 52, 54, 57, 58, 60, 65, 66,
        70
Memory Gate (MG), 214
Metal-Oxide-Metal (MOM) capacitors, 236,
        237

Metal-Oxide-Nitride-Oxide-Silicon (MONOS),
        47, 48, 71
MG decoder, 194, 195–200
Micro-Controller Unit (MCU), 1, 3, 75, 76,
        80–85, 89, 90, 94–96, 98, 99, 102, 103,
        111, 117, 119, 120, 122, 126, 128
Multiple-Time Programmable (MTP) memory,
        31

N
NAND-flash, 3
Nano-crystal/Nano-dot, 42, 43, 47, 48
Nitride film (Si_3N_4), 209–211, 224, 225, 243
Nominal VDD (Vddnom), 166
Non-boosted, 219
Non-volatility, 3
NOR, 75, 77, 85, 87, 118

O
One-Time Programmable (OTP) memory, 7–9,
        30
Optical Process Correction (OPC), 158
Over-erase, 36–38, 40, 41, 46, 48
Oxide-Nitride-Oxide (ONO) film, 214
Oxide-RAM, 3

P
Page flash, 87
Poly-Insulator-Poly (PIP) capacitors, 236, 237
Poly to poly FN electron ejection, 36, 40
Pre-write, 192, 193
Program disturb, 37, 38, 53, 60, 185, 196, 215
Programmability, 3

R
RAM, 7, 21, 23
Random access, 228–230
Read circuit design, 52
Read disturb, 32, 46, 182, 185, 187, 188, 190,
        196, 200
Read disturb free, 188, 189
Read Only Memory (ROM), 7–10, 12, 15, 18,
        20, 21, 23, 131
Real-time, 1–3
Redundancy, 156
Reliability, 29–32, 38, 39, 41, 42, 45, 46, 52,
        53, 60, 62, 65, 67, 70, 165, 179–181,
        201, 206
Retention, 35, 39, 45, 46, 59, 209, 224–226,
        241, 242

S
Safety function, 62, 66, 67, 71

Scalability, 36, 39, 42, 45
Sector erase, 85, 88
Secure, 75, 80–85, 89, 90, 94–96, 98, 99, 102, 120
Secure-MCU, 3, 20
Security, 11, 13, 15–18, 20–22, 25, 147
Security function, 49, 62, 66, 68
Sense amplifier, 217, 219, 227, 228, 243
Sense Amplifier with Digital Offset Cancellation (SA-DOC), 227, 228, 231, 243
Sensing, 161
Sensing scheme, 59, 60
SG-MONOS, 214, 223, 225, 240, 243
Simple structure, 181, 206
Smart card, 20, 150
Smart Clock Generator (SCG), 201–203
SONOS, 23, 42, 46, 48, 71, 179–181, 183, 185, 190
SONOS split-gate, 209–212, 214, 217–219, 221–225, 227, 231, 238, 240–243
Source FN electron ejection, 36
Source-Side Channel Hot Electron (SS CHE), 141
Source Side Injection (SSI), 35, 40, 44, 45, 71, 211, 213, 215, 221, 242
Split-gate, 148
Split-gate cell structure (1.5Tr), 32, 33, 35, 37, 40, 41, 47, 48, 58
Spread Spectrum Clock Generation (SSCG), 237, 238, 243

SRAM, 3
Standard IO-MOS, 195–200
Stress Induced Leakage Current (SILC), 39–41, 45, 140
SuperFlash, 148

T
1T, 111–113, 115–117, 119, 120, 122, 123
TDDB, 229, 240
Thermionic emission, 224, 225
Threshold voltage (Vt), 164
1.5Tr, 209, 211
1 Transistor cell structure (1Tr), 31–33, 36, 37, 40, 41, 46, 48, 51–53, 58, 71
2 Transistor cell structure (2Tr), 32, 33, 37, 40, 41, 48, 50, 58, 71
Trap distribution, 225, 226
1Tr-SONOS, 179–187, 189, 190, 194, 195, 198, 204–207
1Tr. structure, 179, 180, 187, 192

V
Voltage switch (VSW), 54

W
Word Line (WL), 209, 219, 240

X
X-Decoder (XDEC), 158